HOT STEEL

THE STORY OF THE
58TH ARMORED FIELD ARTILLERY BATTALION

EDITED BY

FRAN BAKER

Cover Design and Interior Format

© KILLION
THE
GROUP, INC.

TABLE OF CONTENTS

ACTIVATION

The 58th Field Artillery Battalion (Armored) was activated on 1 October 1941, at Fort Knox, Kentucky. The original personnel were drawn from the 22nd Field Artillery Battalion (Armored), 54th Field Artillery Battalion (Armored), Armored Force School, and Armored Force Replacement Training Center at Fort Knox. The new Battalion, under command of Major John G. Howard, had an original strength of 35 officers and 315 enlisted men.

In infancy, the Battalion was nourished in the confines of Fort Knox's "Tent City", and the only barracks were the well-known pyramidal tents. Winter was cold in Fort Knox, and overheated Sibley stoves were constantly setting tents afire. The earliest recollection of guard duty in the 58th is of the G.I. patrolling the area complete with fire extinguisher and baseball bat.

Eight days after the bombing of Pearl Harbor by the Japanese on 7 December 1941, Battery D of the Battalion was inactivated. Sixteen officers and 107 enlisted men were transferred to the 705th Tank Destroyer Battalion, leaving the 58th with three firing batteries, Headquarters Battery, and Service Battery. On the same date the medical detachment was also activated. Gradual additions to the strength raised it to 48 officers and 473 enlisted men by 1 January 1942. At that time the Battalion was redesignated the 58th Armored Field Artillery Battalion, an organic unit of the 5th Armored Division.

Captain Bernard W. McQuade assumed command of the Battalion on 31 January 1942. On 11 February, the Battalion was split—58 men going to the 6th Armored Division, 124 to the 8th Armored Division, and the balance of 18 officers and 240 enlisted men making a permanent change of station to Camp Cooke, California, arriving on 16 February 1942.

On 10 March 1942, Major Bertram A. Holzworth assumed command of the Battalion. During that month 427 replacements came to the Battalion to receive their basic training on the windswept sands of Camp Cooke, and to be assigned to the Battalion on 18 May 1942. Thus, the Battalion entered its artillery training with a complement of 22 officers and 717 enlisted men.

During the latter part of May the Battalion was alerted. On the 31st, together with certain other elements of the 5th Armored Division, the 58th moved down the coast to the vicinity of Los Angeles to protect the Pacific Coast from the Japanese Navy. No serious development came of this threat, but the Battalion did make its initial step on the climb to the pinnacle of fame when a lowly Service Battery private looked upon the wine when it was red and decided to sleep in the same tent with the

Commanding General of Combat Command A.

On 26 June 1942, Lt. Col. Holzworth was transferred to the 95th Field Artillery, and Major McQuade reassumed command of the Battalion.

Training continued at Camp Cooke until 6 August, when the 5th Armored Division began movement to the California desert for maneuvers. At Freda, California, the Battalion bivouaced and tried to acclimate itself to the withering heat. The first few days in this sun-baked oven saw men collapsing of heat prostration, and the remainder dragging listlessly under temperatures ranging as high as 142 degrees. When the Battalion had struggled through the perspiration and dust of one overnight problem, and was just beginning to feel it could stick it out, orders came down to prepare to move to Camp A.P. Hill, Virginia. By this time we had been detached from the 5th Armored Division and had become Corps troops that were attached for varying periods of time to different divisions or combat teams.

Early in the morning of 7 September the Battalion detrained into the rain and red mud of Ole Virginny. In the wee, small hours, and under a steady downpour, they formed two ranks, paired off, and pitched pup tents. It was rough, men, it was rough! The Battalion remained at A.P. Hill for nearly a month, during which time almost everyone received three-day passes, and everyone got the diarrhea. Here also 18 brand-new Carriage 105 mm howitzer M-7s (SP) were forced upon them. This was the go-ahead signal for Camp Kilmer, N.J. on 11 October. Things were happening fast, and it began to look as though the Battalion was going somewhere besides home on furlough.

At Camp Kilmer it was no longer a secret in the Battalion that the destination was overseas. The Battalion was in "quarantine" now, they were told, as part of the last few preparations for overseas. The mail censor came into the men's lives, with orders that the Battalion location must remain secret and no mention must be made to those "outside" of the impending embarkation.

The first morning at Kilmer ushered in a clothing check—an affair which became more or less a daily routine during the Battalion's stay there. In the morning, first thing, you scattered your belongings over the bed. Each night you packed them carefully away in the barracks bags, and the next morning did it all over again. Everyone had photographs taken for identification cards, but no one ever found out what happened to the pictures.

The personnel section became a madhouse, with last minute allotments, insurance, safe-arrival cards, replacements, and transfers to be taken care of. The safe-arrival cards, to be mailed out to the families when the men arrived at their overseas destination, furnish another minor mystery of the war. Apparently a large part of the Battalion never "safely arrived."

After the Army way, there was a "dry-run" before the real thing came off. Thus, the Battalion loped from the barracks to the railroad station with full field and one barracks bag—just to see how it felt. Similarly, one afternoon the Battalion was lined up and filed through a dummy gangway, in loading order.

The phone booths in the P.X. and the Service Club did a rushing business those last few days. The men could not give any information, but there was an urge to have one more chat with the folks, the wife, the gal friend. When the 58th finally sailed away, they left a sizable wake of collect phone bills. And the packages that went home, clothing and equipment was limited to T/BA requirements, and all those oxfords, extra socks, underwear, etc., were supposed to be sent home, turned, in, or thrown away. In reality, about 10 percent was sent home, 10 percent turned in, 5 percent thrown away, and 75 percent tucked in between blankets and in pockets of overcoats.

Finally, on the morning of 1 November 1942, the Battalion lined up in front of the barracks and trudged off to the station. On the train they sat in the midst of a sea of barracks bags and equipment. A few hours later they dismounted at the terminal and crowded onto the ferry. Going across the harbor to the pier the ferry skirted the Statue of Liberty. It was probably a common thought: "When'll we see you next, Sis?"

SHIP AHOY!

The Battalion walked down under the sheds of the pier and then up the gangway to the deck of the Santa Rosa. Any illusions about traveling first class were quickly dispelled as the men stumbled down to C and D Decks and viewed the hammock-like bunks. The next morning, the men awakened to the gentle roll of the ship. She had shoved off during the night, and the Battalion was on its way—where to, no one had a clue.

Appetites were still fairly hearty, and the men converged on the mess hall on A Deck. There they filed through what had apparently once been the ship's ball room, past an array of steam tables. Five feet inside the hall the men lost their hearty appetites, not to regain them the remainder of the trip. They tripped gaily and innocently through the doors, and there the essence of steamed scrambled eggs of dubious vintage smote them like a whiplash. It was a terrible let-down, but the men of the 58th were soldiers. They came back again and again, pale but determined. It seems not possible that the field of battle could hold terrors for men who have survived the chow line aboard the Santa Rosa.

Shortly after breakfast the men were allowed up on deck, and many of them spent the greater part of their time up there during the trip. Each man was given a lifebelt, or jacket rather, to be worn at all times. They were cautioned to sleep with clothing on, "just in case." Inasmuch as they were wearing "long-handled" drawers at the time, this became rather unpleasant. As a result, each night the men turned in displaying various forms of undress.

Before long, symptoms of "mal-de-mer" made themselves apparent. The seas were not unusually rough, but there is something about the steady pitch and toss of ordinary ground swells that eventually got to some of the men. A large garbage can was placed in the middle of the sleeping quarters, forward on C Deck, for use in disposing of trash. It was quickly converted into a gathering place for those unfortunates whose insides revolted under the strain. It was quite common, in fact, to see three or four dogfaces clinging grimly to the edges of the can and giving it all they had.

If you found a group with heads together, with no can in the middle, you could safely conclude that it was a poker or crap game. That was about the limit of the recreational facilities on board, and the men made the most of it. Some big games developed. All day long a number of sets of ivory cubes would keep galloping up and down the deck. By the time the Battalion got off the boat, some of the boys had paid for a handsome first class passage, while others landed with small fortunes.

A few days out, the anti-aircraft crews started practicing periodically. They opened up with 40 mm and .50 Caliber guns, to send tracer arcing out over the water. These occasional outbursts, coupled with two or three accidental discharges of one of the bow guns, encouraged the changing of the ship's name from "Santa Rosa" to "Roaring Rosa."

The ship carried a large stock of P.X. supplies, especially candy. The men consumed an ungodly amount of Hershey bars, Baby Ruth bars, and soft drinks. Today the remaining men in the Battalion swear to high heaven that chocolate bars kept them alive crossing the Atlantic. Going through the chow line was a mere test of courage and fortitude—they got their nourishment in cartons from the P.X.

A couple days of comparatively rough weather were experienced. The ship developed quite a roll. Captain Brewer mounted a pendulum on the wall of his stateroom with a scale to register the degree of the roll. After every particularly violent lurch, officers would pour into the stateroom to ascertain whether or not a new high had been recorded. The mess hall was really a mess on those days. In addition to the usual risks confronting the hardy wayfarer into the chow line, the deck was coated with gravy. When maneuvering in the open, away from supporting tables, it was best to maintain a slight crouch with knees slightly bent and body well forward, as in skiing. Sun Valley never produced a downhill slalom run like the course from the last steam table to the slop bucket on the Santa Rosa on a rough day.

When about six days out, the men were notified that their destination was the West Coast of Africa. An invasion force was already striking at the continent at several points. The appointed landing place for the Battalion was to be Casablanca, but that would have to depend on the success of the initial invasion forces. There was a possibility, they were told, that they might have to land under fire themselves. So (yes, you guessed it), a dry run was held on landing operations. Large cargo nets, for use in clambering over the side of the ship, were hung on one of the hatches between C and D decks, and one fine morning the men donned full field, hauled themselves up narrow companionways, out on deck, and down again to C deck, where they dropped, one by one, down the net to the lower deck.

Daily estimates as to the position of the ship ran riot. Rumor had it spotted all the way from the coast of Portugal to the Cape of Good Hope. A wall map on B deck was so well gone over that by the time the ship was ten days out the whole eastern Atlantic was one grease spot. In the early afternoon of 17 November the first blurry haze on the horizon, which rapidly developed into land, was picked up. This trip of the Santa Rosa knocked two full days off Columbus's best time for the same crossing. From the deck of the ship the green and white beauty of Casablanca was a most welcome sight. From a distance it was clean and fresh and cool.

Late that afternoon the ship felt its way into the harbor of Casablanca, while the destroyers paced back and forth just outside. There was no question about it: the war had visited the Casablanca waterfront. Freighters and passenger liners were heeled over or resting on the shallow bottom with only the rigging showing above the water. A cruiser and destroyer lay half-submerged on the far fringe of the harbor. Wreckage floated on the water.

As we were warped along through the debris and up to the dock, the men crowded the rails and stared at their first eyeful of the wake of battle. On the far side of the pier could be seen the superstructure of the huge French battleship Jean Bart, settled on the bottom of her berth where American bombers had found her on the 9th, helpless, not yet ready for the sea.

At about 8:00 A.M. the next morning, 18 November, the men crawled into full field, slung "A" barracks bags over their shoulders and marched off the Santa Rosa. To most of the men that first step off the gangplank was their first step on foreign soil. They were invaders leading the way for America's long awaited "push." But you can't think of those things when all your soul and being is concentrated on keeping your barracks bag from rolling around your back and decapitating you.

14

So the big moment went by with no contemplation on its significance. The 58th probably looked more like a gang of stevedores at Memphis than a conquering army at Casablanca.

NORTH AFRICA

Once off the ship the men lined up, readjusted their burdens, and took off in as close to a march formation as could be managed under the circumstances. The physical condition of the Battalion, after the long stay on ship, was not the best, and the half-mile they carried the barracks bags was long and torturous. When they finally reached the dumping place for the bags, some of them were dragging on the ground. And that's not all that was dragging!

With the disposal of the bags, morale rose sharply and the march to the bivouac area was resumed. There were no bands or throngs of people throwing roses in our path, but the atmosphere was decidedly friendly. It was soon apparent that the city was not as clean and white as it had appeared from the sea, but the long trip was over and the Battalion was safely landed on good, solid terra firma.

The men had on long woolies, the day was warm, and it was with a definite feeling of relief that they reached the bivouac area, some six miles from town, and sat down to eat a K ration. It was their first K ration, and it tasted like the dinner your mother always fixes on your birthday. After the cuisine of the Santa Rosa, that tin of cheese and those biscuits were delicious.

The new bivouac area, unromantically enough, was an Arab cow pasture. No rolling dunes, with desert sheiks riding out of the sunset caroling the "Desert Song." No camel caravans trekking across the wastes. Just an evacuated heifer haven, with a few of the dispossessed moodily browsing on the outskirts. But, home it was to be, and the men set to work pitching tents. Simultaneously with this the rains came. It rained, it rained, and it rained. In the midst of the downpour the barracks bags came up by Arab Express. Next morning all the equipment was a soggy mess, and the next few days were spent trying to dry out the stuff between showers. The large portion of the Battalion's memory of Casablanca is reflected in little pools of muddy water, as well as incidents like the time Sgt. Hopkins called the CP with information that a supply ship had been torpedoed in the battery street.

After a few days spent pulling themselves together and adjusting living quarters, the men fell into the familiar routine of training, calisthenics, and classes. Road marches became the order of the day, and life somewhat resumed its normal course. The big social event was the first pass—a chance to roam the streets of Casablanca and peek into the mysteries of North Africa as well as a chance to bend the elbow at a few bars. We walked into town on the first trip in our wrinkled O.D.s drawn from the recesses of our bags.

17

Casablanca was a rather large city, with fairly modern sections. As in most of North Africa, French was as widely spoken as Arabic and the dominant European element was French. The language barrier offered some obstacle to us at first, but by signs and trial and error, we more or less surmounted that in short order.

The exchange rate was 75 francs to the dollar (later changed to 50 francs). At first prices were quite reasonable, and the American dollar went a long way. However, the Arab merchants and vendors soon awoke to the gold mine which was being visited upon them and raised their prices accordingly. The principal merchandise in the shops was leather goods, for which Morocco is well known. Portfolios, purses, hand bags, billfolds, slippers of intricate design were on display in about every fourth show window in town. Liquid refreshment consisted mainly of beer and red wine. The beer was fair, but the bulk of the wine was not too good—mainly because the average American is apparently not accustomed to the sour red wine of the French. The restaurants boasted a very limited bill-of-fare, leaning strongly to egg omelets, fish, and cabbage soup. The pastry shops received a big play from the soldiers. Most of their products, however, suffered from a lack of sugar and flavoring.

We became quite conscious of the housing problem during our stay at Casablanca. The men started out with simple pup-tens pitched just as they had always been trained to pitch them. The officers got the idea first of bettering their standard of living, and shortly the officer's quarters began to sport rock and board walls for the tents, raised platforms for the beds, etc. Captain Paton's imposing residence was promptly christened "Paton's Medina." As more lumber was obtained the men gradually added to their tents and dug down the floors. Hay had been obtained for mattresses, and the boys got along quite well in this respect. Each tent was surrounded by a miniature moat against the occasional cloud-burst.

The Arab children, and even grown-ups, had developed a passionate fondness for chewing gun, chocolate, cigarettes, and our hard candy from the rations. Everywhere we went we were greeted with cries of "Cig'ret? Choonbomb? Bon-Bon?" All the way across North Africa the same war-cry was heard.

The morning of 31 December Jerry visited Casablanca with a few planes—no one seemed to have a definite idea of how many. About three in the morning the men awoke to what sounded like a general invasion. They piled out of their tents to find the sky a network of tracer, with searchlights crawling and probing, ack-ack flashing here and there. As they watched a bomber was caught by the lights, and then things really cut loose. Everything from .30 calibers to 90 mm's was thrown at it. The Battalion yelled and cheered till it resembled a Saturday afternoon at Ebbetts Field.

Eventually things quieted down, and we retired to our tents.

A few minutes later the anti-aircraft cut loose again, and we came out and rooted a little longer. It was all very exciting, and provided a lovely night spectacle. Later, in an olive grove at Maknassy we were to remember it. At any rate the next morning the Battalion dug slit trenches.

The Battalion was selected to furnish honor guards for several occasions, visits of French Generals, Admirals, etc. The climax of the honor guard business came with the visit of President Franklin Roosevelt and Prime Minister Winston Churchill at Casablanca. Perhaps it couldn't properly be called an "honor guard," but a detail from the Battalion patrolled the road down which the President and his party drove.

The Battalion also furnished unloading details for the dock during the first few weeks, when supplies and equipment were coming in. A few of the tents broke out with unheard of rations shortly after these details. Apparently a few of the lads had made the best of their opportunities.

Usually the day started out with a half-mile run before breakfast , and each morning the chill dawn echoed to the clop, clop, clop of the G.I. shoes on the macadam road skirting the Battalion area. The object was to enable each man to run a mile in 10 minutes with full field. A mile in ten minutes is not such a rugged undertaking, but with full field and rifle or tommy gun slapping you on the behind it is not such a pushover. Headquarters Battery still remembers the morning Captain Paton led his boys double-timing in full field down the road with Sergeant Parker of Service Battery trumpeting out "A-hunting We Will Go" on his bugle.

Christmas 1942 in Casablanca was more or less limited to memories of other Christmases, and hopes for those to come. A clump of sage-like bush, placed outside the orderly room, bravely tried to carry on and perform an unaccustomed duty as a Christmas tree. An engineer outfit from down the road came up on Christmas Eve and staged a tableau of the Nativity. Their costumes were very impromptu, and the outstanding character was a very unwilling donkey. But it did bring a little more Christmas down around the men.

After a three month stay at Casablanca, the Battalion "pulled their stakes" on 16 February 1943 and took the first step north to Rabat—a few kilometers closer to the battlefront. It was a shame to leave all those fine dwellings—our own dugouts as well as the stone and cement foundations on Nob Hill (Officer's Quarters).

The 58th had been assigned to 5th Field Artillery Group on 25 January, along with the 62nd Armored Field Artillery Battalion, and the 65th Armored Field Artillery Battalion. Colonel Newton W. Jones commanded the group. About ten miles out of Rabat the Battalion went into a bivouac area in a cork forest. The cork forest was quite a change from the Arab cow pasture.

In case there is any doubt, corks do not grow on trees like walnuts. The bark is the cork and is cut and peeled off the trunk of the tree.

In accordance with 58th tradition it rained the day the Battalion got there, but after that the weather was increasingly warm. There was plenty of shade. The only ones not happy about the situation were the guards, who had a bit of trouble getting oriented in the woodland on dark nights. After a fight talk from the O.D. at guard mount, the guards would come out with blood in their eye, and it could be marked as a good night to remain in the sanctuary of your tent after dark.

Meanwhile, the German General Rommel was being chased across Tunisia by the combined English and American troops. So on 7 March, the Battalion left Rabat and began movement to the front. Two trains left on the 7th, two trains and the Battalion Commander and convoy of wheeled vehicles on the 8th, and one train on 9 March. The track vehicles were loaded on flat cars, and the crew rode right along in the vehicle. Some of the men could sleep in box cars, but there had to be a guard on the vehicle at all times. The view from the flat cars was much better, so the vehicles were always crowded during the daytime.

Shortly after crossing into Algeria we encountered hills and mountains. And with them came the tunnels. Tunnels in the United States are more or less commonplace, and the traveler takes them in stride. But traveling through tunnels in the Atlas Mountains on a flat car on a French train is something of a novelty. Approaching the longer tunnels the engine would bound eagerly up to the entrance like a terrier running a rat into a hole. A quarter of the way through it would begin to proceed more laboriously and cautiously, until at about the half-way mark it would shuffle to a halt and take a breather, while the men sat in the darkness surrounded by soft coal smoke and steam and bitterly cursing the engineer, North African railways, mountains, and life in general. When the cars finally emerged into light once more, the machine guns were nicely coated with steam rust while everything and everyone else wore a cloak of soot and cinders.

The scenery increased in beauty going through the mountains. The Atlas ranges furnished some grand views, and were especially welcome after the flat, comparatively desolate country of Morocco. The grades often slowed down the train, until the little Arabs could run alongside for long distances, imploring the men to pass out the bon-bons.

The last trainload of half-tracks and M-7s closed into the bivouac area in the hills above Tebessa, Algeria on 17 March. It started to rain that night, of course, and the Battalion arose the next morning in a steady downpour. Orders were received to move the Battalion to Feriana, and shortly before noon the column started out, down the mountain, past Tebessa and on the road toward Feriana. A bivouac north of Feriana was reached that afternoon. The next day, 19 March, orders were received to join the 1st Armored Division. At this time our forces were just beginning to push back after stopping the vicious drive of Rommel's Afrika Corps in the vicinity of Tebessa and Gafsa.

Shortly before midnight on the 19th, the Battalion contacted the 1st Armored Division at Madene El Feds, Tunisia, and at 0215 on the morning of the 20th began movement into the first combat position. Hence, the first M-7s to be used by American troops lurched and growled their way through the Tunisian night, while the crews awaited the morning and the first taste of fight. At 0600 the guns moved into position, and at 0715 the Battalion fired its first combat rounds, registering on a hill crest which later proved to have been an enemy Observation Post (OP).

One other position was occupied that morning. In the afternoon the first enemy planes appeared high overhead, engaged in a dog fight. Also in late afternoon the first prisoners (Italians) passed through the area, in very good spirits. At midnight the Battalion arrived in position 4 miles east of Sened Station. In the morning of the 21st, at a position 2 miles east of Sened Station, an enemy machine gun nest and tanks were brought under fire by the Battalion. Enemy planes swooped over the position, but appeared to be traveling too fast, or too interested elsewhere to bother with it.

Our first combat mission was about to begin. Early in the morning of 22 March, we moved towards Maknassy with orders to hold the west flank of Rommel's forces while they retreated northward past us and other allied troops. Up to this point the Battalion had been breezing along, firing occasionally, with no return fire, and it seemed more like a service practice or maneuver than actual battle. At Maknassy the enemy took the time and trouble to disillusion the boys on this point.

Enemy planes appeared over the Battalion area, and bombed nearby points in the morning and afternoon. One bomb fell in B Battery area. The Battalion greeted the planes with a creditable display of fire from the .50 and .30 Caliber machine guns. There was a lot of noise, but apparently no casualties took place on either side. On the afternoon of the 22nd a reconnaissance party from the Battalion was shelled while selecting a new position, and the Battalion Commander's half-track strafed and shot-up by enemy planes—again, without casualties.

Late that night a position was occupied just east of Maknassy, in an olive grove. (The Battalion was to become quite familiar with olive groves in the next few weeks; the olive branch may be a symbol of peace, but not with a 105mm howitzer sticking out from behind it.) The next morning when the Messerschmitts came over, one of them was struck in the bomb rack by anti-aircraft fire. One wing came off, and the ship seemed to come apart in mid-air. There were pieces of Messerschmitt all over the area.

On 24 March the Battalion moved out of the olive grove, into the open, but that didn't last long. The position was shelled continuously during the 25th and two M-7s were lost. That evening the 58th staged a "strategic withdrawal" to the confines of the olive grove.

From 25 March to 9 April the Battalion remained in the olive grove. Enemy artillery had found us by now, and the 58th was serenaded frequently by choirs of 88mm guns, of which the Germans seemed to have quite a few. A number of 88 shells were thrown into the fire direction area one Sunday, forcing that section to change its position. Apparently they got out in good time, because Jerry really laid them in just after they left. A large percentage of the shells were either duds or armor-piercing, as they failed to explode upon landing.

The rising scream of the 88 shell in full flight became quite familiar and was dubbed "Whistling Dick." First would be heard the dull "Boom, Boom, Boom" in rapid succession. Then "Wheeeee! Wheeeee! Wheeeee!" and "WHAM! WHAM! WHAM!" These were always accompanied by a sprint and dive for the slit trench.

Someone has said that the greatest invention of the war was the slit-trench. It is one piece of work undertaken and accomplished without urging. A slit-trench is like

26

money in the bank—something on which to fall back when the going gets rough. It is a cozy nook to which the soldier may retire hastily when the cares of the day (and the 88 shells) close in around him. On occasion it can become his temple and house of prayer. If you don't think so, just try lying in one while that whistle shrieks that it's going to crawl right in there with you. The much-discussed democracy of the American Army reaches its zenith in a slit-trench, and there is always room for one more.

Another incident at Maknassy proved that the tanks carrying our howitzers could burn. As we were firing nearly at our maximum range we were ordered to move out of our protective olive grove onto the bare plain. The Germans soon found us and zeroed in with their 88s. One of their rounds hit one of our gun carriers and ignited the folded-up camouflage net sitting on it. The attempt by the crew chief to put out the fire with a hand extinguisher proved futile, the fire got worse, and the vehicle had to be abandoned. It burned on the plain for many hours, exploding ammunition and gasoline. Eventually the gun and carrier were completely destroyed.

Sgt. James C. Hipp remarked long afterward that the Battalion had 12 Stukas attached to it during the stay at Maknassy. They came over sharply on the hour three times a day. And occasionally the night shift would go up hanging parachute flare "like boulevard lights," searching for a likely target in the resulting illumination. Their bombing, however, was surprisingly inaccurate.

It was warm, quite warm, in the daytime now. And cold, oh, so cold, at night. There was always vehicle guard and a couple hours a night of shivering wakefulness beside the vehicle. At night spasmodic fire from the front lines could be heard—the American and German machine guns laying a tenor and alto duet in the dark hours just before the steady rumbling roll of mortar fire. The Lord deliver us from the Infantry and a shallow slit-trench with the enemy throwing everything but the kitchen sink.

The men were eating chiefly C rations now, as they did all through the campaign, though in its later stages the kitchen trucks came up quite often. Appetites were pretty well maintained—though the shiny little tin cans of stew and hash as well as the hard candy that came with them were bitterly cursed.

During this period the Battalion was in support of the 60th Infantry Regiment, 1st Armored Division. On 1 April word was received that Major McQuade was promoted to Lt. Colonel, effective 20 March. Eventually Rommel and the Germans retreated to the north and we were no longer needed at Maknassy. So on 9 April the Battalion departed from the olive grove in a cloud of dust.

With buoyant heart and armored chassis,
We arrived before Maknassy,
Where the 88s sleek and sassy
Spoke to us in accent brassy.
Bid us pause at old Maknassy,
The situation reached impasse,
And we stayed at old Maknassy;
Far from stream or hillside grassy,
Far from home and fair-skinned lassy.
We got damned sick of old Maknassy,
We got damned sick of old Maknassy.

After a 68-mile, dusty, all night march, the Battalion arrived at Bir El Hafey, where equipment and ordinance were checked. The next area, Sadaguia, was reached the afternoon of 12 April. From here the Battalion traveled 208 miles to La Calle, Algeria, arriving 16 April. The Mareth Lane had been cracked by the British 8th Army, and American forces had completed their junction with the 8th. Rommel was being crowded now—and crowded fast.

At La Calle the Battalion was attached to the 13th FA Brigade, and on 21 April departed for Beja, Tunisia, some 69 miles away. The Battalion approached Beja simultaneously with spring. The hillsides were beginning to bloom, and the column passed through entire valleys of crimson poppies spilling down the hills and across the fields. Solid patches of powder blue and gold carpeted the roadside. A number of German tanks were also spilled across the fields in this sector—Mark IVs and Mark VIs. From Beja the Battalion went into position in the vicinity of Sidi Rovine near midnight of the 23rd, high in the hills.

From 27 April to 4 May, the Battalion was in position at Djebel Zeraris, firing intermittently. Enemy planes again came over the position and bombed in the valley below. A few rounds of enemy artillery came over, but it was neither consistent nor accurate.

On 4 May the Battalion moved approximately 12 miles to a position near Eddekhelia (Dhekelia). In the early hours of 5 May, the Battalion then joined in a tremendous barrage laid down on the now historic "Hill 609." Enemy artillery fire was falling in the Battalion area during the day and the road was under fire, but no damage was done to the Battalion. At this point the Battalion was supporting the 168th Infantry of the 34th Infantry Division.

On 6 May the Battalion was attached to the 1st Armored Division and moved to the vicinity of Mateur. The reconnoitered position on a hillside above Mateur was already occupied by enemy tanks when we moved up. A Battery opened up with direct fire, destroying one tank. Darkness fell with the hilltop still in German possession, tanks burning on the hillside, tracers arcing up from our positions at the foot of the hill, and the Battalion firing time fire on the hill crest.

On the morning of 7 May, just after dawn, enemy artillery fire began falling in the Battalion area. Captain Kenneth Bitler of A Battery could see the flashes from the

German guns. Observing his own fire, he zeroed in on the enemy guns and soon silenced them. The Battalion then took the position reconnoitered the previous day, and fired on targets in the vicinity of Ferryville.

On the afternoon of 7 May the Battalion moved to a cross-road two miles north of Ferryville, and fired through the night on enemy tank concentrations. The morning of 8 May the Battalion was moved again through the cheering throngs of Ferryville, to a position five miles east of the town. While in this position, one of the Battalion's observers, 1st Lt. Carl M. Johnstone, observing from a hill overlooking a large valley 10 miles east of Ferryville, found a large concentration of enemy armor and mobile field guns being displaced in an attempt to stop the advance of the 1st Armored Division. He directed fires of the Battalion on this concentration for several hours, forcing the enemy vehicles out into the open where our tanks and TDs were able to destroy them.

With the front lines advancing rapidly, the Battalion moved again in the evening of 8 May and went into position in support of the 13th Armored Regiment. The war had now developed into a field day for our observers, and numerous lucrative targets were fired on by the Battalion. That night large fires burned on the horizon, an indication that the trapped German forces were destroying their equipment.

On the morning of 9 May our front line

elements were pressing toward Bizerte with little or no resistance. The enemy was beginning to give up, and the Allied forces were soon overwhelmed with prisoners of war and captured enemy and civilian vehicles. Finally, at 1121 hours on 9 May 1943, the message was received from Division Artillery to "Cease firing in all present positions. The Germans have surrendered."

The Battalion moved again in the afternoon to a final position on the Bizerte-Tunis road. One lone German artillery piece sounded off as the column moved up to the position, but the gunner must have seen the light very shortly, and quiet reigned on the Tunisian front. The rest of that day, surrendering German troops moved in procession down out of the hills and wadis. The war in Africa was over.

From 11 to 17 May the Battalion remained in bivouac four miles southeast of Mateur, cleaning up, resting, and a few men taking passes to Tunis. On the 18th the Battalion began movement to a bivouac area at Gastu, Algeria, part going by road and part by water from Bizerte to Phillipeville. Until 15 June, the Battalion remained in bivouac at Gastu, resting and performing maintenance. Passes were given to Bone, Phillipeville, and Constantine during this period. On 16 June the Battalion moved 210 miles to El Alia, Tunisia, where it remained until 7 July, preparing for the invasion of Sicily.

SICILY

For the Sicilian invasion, the Battalion was divided into two groups. The first group consisted of a cut-down Headquarters, three firing Batteries (whose guns were taken over by 3rd Division Artillery), three radio half-tracks, ammunition vehicles and a service element. The second group contained the balance of the Battalion. On 12 July, the first group, or first follow-up, loaded on LST-351 in the harbor of Bizerte. The following day, the 13th, the LST pulled away from the dock and joined the convoy proceeding toward Sicily.

The southern coast of Sicily had been secured by other forces when we landed at Licata, Sicily. Waterproofing was removed from the vehicles at an area five miles north of Licata, and the Battalion began the march through the hills to the vicinity of Riesi to support the 3rd Battalion, 30th Infantry, of the 3rd Infantry Division. The move was completed near midnight and parties were sent out to pick up the Battalion's M-7s which had been taken in on the invasion by the 10th, 39th, and 41st Field Artillery Battalions.

The following day, 15 July, all the M-7s reached the Battalion area. One mission was fired during the day, and two or three enemy rounds whistled in, but not too

33

close. On 16 July, orders were received from Division Artillery to move to an assembly area in the vicinity of Favara. We reached this position in mid-afternoon, and shortly thereafter the Battalion fired on enemy infantry and vehicles, causing many enemy casualties, and forcing enemy infantry to come out in the open and surrender. This fire continued on the morning of the 17th, and in late afternoon the Battalion moved to a position five miles north of Canicatti, in support of the 15th Infantry.

On 18 July the Battalion moved again to Serradifalco, to find that all resistance in the area had collapsed, and at midnight was on the move again to a new position. This constant moving continued through the first days of the campaign, until the Battalion ran out of gas on 21 July, near Corleone, while still supporting the 15th Infantry. From Serradifalco to Milena, to Station De Sutera, to Mussomeli, Aqua Viva, Castianova, Priss and Corleone, the opposition consisted mainly of Italians, and the resistance they offered was light as they frequently ran out in the open and surrendered after two or three rounds from our guns.

The terrain was rugged. Long night marches over the torturous climbing trail-roads of Sicily demanded every last ounce of stamina of the drivers. The Battalion moved from village to village, scarcely pausing long enough to place the guns in position.

On 22 July the 58th Armored Field

34

Artillery Battalion was given the mission by the 3rd Infantry Division Commander of driving to the coast, seizing the seacoast town of Trabia, and cutting off the enemy retreat from Palermo to the east. At noon that day the Battalion began the wild ride, which took it through Godrano, where 200 prisoners were captured by the Battalion; Cefala, where 200 more prisoners were taken; and then through Villafrati, Raucina, and Ventimiglia. Out of Ventimiglia a bridge had just been blown by the retreating enemy, halting the push while a party from the Battalion devised an impromptu by-pass of the blown bridge so the Battalion could continue on its way.

At 1740 hours advance elements of the Battalion entered Trabia. There was no resistance, and 1,811 prisoners were taken. The retreating enemy had planted mines under the road in the town, but these were removed almost voluntarily by the Italian prisoners. A Battalion patrol entered Altavilla at 2030 hours, taking 210 prisoners.

The next day the Battalion occupied Termini Imerse. The commander of the Coast Defense Section surrendered to the Battalion CO, and a battalion of Italian Mountain Artillery surrendered to the 58th. The Battalion set up civil government in Trabia, Termini Imerse, and Caccomo. The old Five-Eight was really taking over. A total of 3060 prisoners had been captured, and three sea-coast towns taken without the support of infantry or tanks. For 24 hours the 58th operated a

Prisoner of War camp on the hillside above Trabia that fairly teemed with milling Italians. Then the AMGOT and 30th Infantry Regiment took over.

On 26 July the Battalion marched out of Termini Imerse and up into the hills to Petralia, where the 5th Armored Artillery Group was attached to II Corps. The second follow-up of the Battalion arrived in the area shortly before the first follow-up closed, and once again the Battalion was consolidated. The Battalion remained in this area until 30 July and then moved to an assembly area near Sperlinga. Here orders came up assigning the Battalion in support of the 179th Infantry of the 45th Division, which was moving on Mistretta. At 2145 the Battalion moved into position west of Mistretta, just forward of the infantry front lines.

We remained in this position until 3 August. Blown bridges, which the rugged terrain made impossible to by-pass, prevented vehicles from moving on San Stefano, on the coast. After reconnaissance established that the move could be made on 3 August, the Battalion moved at 2200 hours down from the hills and up the coast in the vicinity of San Stefano, pulling into position at 0300 hours the morning of 4 August.

The Battalion was now attached to the 3rd Infantry Division. That afternoon the ammunition trailers had to be left out by the main road, and the guns had struggle enough grinding up over the rocky trail to get into position. No sooner was the position occupied than enemy artillery began pouring in. However defilade was excellent and a mask of trees in front of the Battalion position saved it from direct hits. But the Germans kept on trying, and the 58th sat it out and watched the bursts behind them. Some even went out to sea and burst in the water of the Mediterranean. The shelling continued with the same effect throughout the 5th and into the 6th. On the 5th one man in A Battery was killed by a time burst (the first man killed in the Battalion by enemy action) and one other wounded.

At noon on 6 August, the Battalion pulled back into the previous position. The Germans still held the east part of the island, and higher command drew up a plan to make an end run by sea and to land an amphibious force behind enemy lines along the north shore. A and B Batteries were selected to be part of this operation in support of the 2nd Battalion, 30th Infantry Regiment.

On the evening of 7 August, the two batteries, with 4 guns each, loaded on landing craft in the vicinity of San Stefano. Two observers from Headquarters Battery were attached. The landing was made at 0345 hours on the morning of 8 August, in the vicinity of San Agata. Slight opposition was encountered and the landing was a complete success. The remainder of the Battalion moved forward to join A and

B Batteries, arriving in position near San Agata early in the morning of 9 August.

Throughout the night persistent enemy counter-battery fell in the area. Three men were wounded during the day, but no other damage was inflicted. On 10 August, a second amphibious operation was planned with the same batteries and personnel participating in support of the 2nd Battalion, 30th Infantry Regiment. At 0335 hours that morning the landing was made near Brolo, Sicily. And on this day the luck of the 58th ran out. There was no surprise this time. The Germans were waiting on the shore, and 20 mm fire greeted the landing parties before they reached shore. In addition, a battalion of German infantry, anti-tank guns and a half dozen Mark VI and Mark IV tanks resisted the landing. The quarter-ton vehicles and ammunition trailers couldn't get off the beach, and had to be abandoned there.

The first position occupied was just northwest of the town of Brolo. The trees in front of the guns were so close that only Charge 1 could be used in firing on the town. The guns were moved west to permit use of Charge 7. Enemy tanks on the road then destroyed two of A Battery's M-7s. And from then on the Germans threw everything but the well-known kitchen sink at the landing party. B Battery lost three M-7s—two by enemy fire, and one apparently through mistaken bombing by friendly planes.

One M-7 from each battery reached the

main coastal road and took up positions to fire on enemy tanks. Both were shortly destroyed by enemy fire. Enemy artillery, mortar, and machine gun fire was now constantly sweeping the area. When all the M-7s were destroyed the men joined the infantry, which was taking a position on the hill across the road from the beach.

The Battalion Executive Officer, Major Stuart B. Lamkin, was killed while firing a machine gun to cover the withdrawal up the hill of his men. 1st Lt. William J. Murray, forward observer (FO), was also killed while attempting to direct fire. The Commander of A Battery, Captain Richard M. Rossbach, and his Executive, 1st Lt. Martin J. Keiser, were both captured. The total count of losses for the action revealed 9 killed, 14 captured, and 25 wounded. Seven out of the eight M-7s taken on the landing were totally destroyed, as well as one half-track. On 12 August, infantry moving up the coast broke through to the battered landing party, and the survivors of the Battalion moved back to the position at San Agata.

Thus the 58th's war in Sicily came to a close. The actual end of resistance on the island came a few days later, but for the Battalion it ended with Brolo.

The Battalion remained in bivouac in the lemon grove at San Agata until 24 August. The next move was made in the early morning hours, in the first real rainstorm encountered in Sicily. Throughout the series of campaigns, it had become accepted tradition that the 58th never moves until it rains like hell. The new bivouac area was near Termici, Sicily, and was reached after some 10 hours driving down the coast.

The war being over—temporarily, at least—passes to Palermo were in effect once more. Also, some enterprising soul found a mountain in the back yard which we could climb, just to keep in shape. Many of the men swam in the blue waters of the Mediterranean—than which there is no bluer.

On 10 September, the Battalion moved back through Palermo to a bivouac area between Trabia and Termini Imerse, the towns the 58th had liberated during the campaigns, and therefore more or less considered "their towns." The Battalion pulled into the area with instructions not to unpack too much, as it was not anticipated they would remain long. Loading and personnel lists were there prepared, and it looked as though the 58th was Italy bound. However, on 15 September, orders were received relieving the Battalion from attachment to the 7th Army and from assignment to the 5th Army.

Until 14 November the 58th remained in bivouac in the olive grove overlooking the Mediterranean. Passes continued to Palermo, the city with the blasted waterfront. A pass in Palermo consisted chiefly of roaming the streets in search of *vino*, and being accosted by numerous individuals who yelped "Spaghetti, Joe?" "Chicken, Joe?" at you. To prevent the Battalion from succumbing to the lassitude of the late Sicilian summer, a convenient mountain was selected and scampered up, and calisthenics were taken every morning, during which the 2nd Lieutenants took turns trying to wear out their batteries.

General George S. Patton, Jr. visited the area of Fifth Group to review the troops and present individual awards won during the Sicilian Campaign. Silver Star awards were presented to several members of the 58th.

On 12 November the Battalion turned in all vehicles, and on the 14th moved to a staging area at Mondello Beach, six miles west of Palermo. Here, at another ceremony, General Patton presented two Distinguished Service Crosses and five Legion of Merit awards to the 58th. On 17 November the Battalion moved to the docks at Palermo and loaded on the British transport HMS Aorangi. On the morning of the 18th, the Aorangi pulled out from the harbor of Palermo, England bound.

The trip to England was another prolonged voyage. The ship docked at Oran, North Africa on 22 November, and remained there five days. It was discovered long after the Battalion had arrived in England that President Roosevelt had timed his visit to Oran to coincide with that of the 58th. Unfortunately he was unable to get together with us.

From Oran, the Aorangi moved to Gibraltar, the celebrated Bastion of England and the Prudential Insurance Company. Two days were spent anchored off Gibraltar, then on to the British Isles. Though the trip was quiet, it was generally known that a German "Submarine Pack" was on the prowl off the coast of Ireland at this time. Radio reports confirmed three subs sunk one day and six another by Allied planes and destroyers.

The Battalion spent Thanksgiving Day aboard His Majesty's Transport Aorangi, and actually had turkey at a delayed Thanksgiving Dinner. They also had all the tea they could drink and all the steamed fish they could eat—and a little they couldn't eat. There was a special Mess for NCOs of the first three grades, an institution which strained old friendships to the breaking point. However, all's well that ends well. The Aorangi didn't sink, as was frequently predicted during the voyage, and the Battalion landed at Glasgow, Scotland, in good condition on 9 December. Debarking at night, the men of the Battalion filed through the long station, where they were welcomed to the British Isles by Red Cross girls with coffee and doughnuts. From Glasgow, the 58th entrained for England.

ENGLAND

Here begins the most delightful period of the 58th's stay overseas. The Battalion was billeted at Adderbury House, near Oxford. Enlisted men lived in Nissen huts with straw-stuffed mattress bags, and the officers occupied the manor house. For the men in the huts, it wasn't exactly all the comforts of home, but after sleeping in the open for a year, it was strictly up-town. It was grand to be among English-speaking people again, and the residents of Adderbury accepted us with smiles and cheer. They even permitted us to blast the quiet, neighborly atmosphere of their pubs, standing aside with quizzical smiles as the G.I.s rushed the bar.

Christmas 1943 in England was vastly different from Christmas 1942 in Casablanca. Although celebration and feasting was curtailed in war-time England, the people invited us into their homes, and the real Christmas spirit was there. (There was also a limited ration of Scotch at the pubs, but of course that's neither here nor there.) The men of the Battalion gave a Christmas party for the children of Adderbury, contributing sweets from their rations and from packages from home. Sgt. John R. Jackson presided as Santa Claus, and looked more like Santa than Santa.

43

Five 2nd Lieutenants joined the Battalion on 3 January 1944, and several other officers transferred to other groups or divisions. The officers hosted a tea for the townspeople and, later, a dance at the house. Each battery had its own dances, at which there was always a goodly attendance of the local belles. A number of strong romantic attachments were developed, and for long after, the Battalion mail call from England ran a close second to the one from home. The Battalion operations sergeant almost had to be dragged out of Adderbury in chains when we left, and he still swears he's going back to marry the gal.

But life was not all tea and roses. War loomed on the horizon. Lt. Kenneth M. Frawley conducted classes in aircraft identification four hours per day for eight men per battery. Lt. Col. McQuade, Battalion CO, and Capt. Bradford Smith, as S-3, gave a lecture to the 87th Field Artillery on the 58th's combat experiences. And Lt. Arthur Rosenbaum, S-2, spent three weeks at intelligence school, returning to the Battalion on 20 January. Then on 28 February the entire Battalion moved to Ivybridge to join the 29th Infantry Division and begin training for the invasion of France.

On 2 March, elements of the Battalion moved to Camp D-19, in the vicinity of Dorchester, to participate in Problem "Fox," with the 116th Combat Team of the 29th. From Camp D-10, a cold wind-swept stretch of woods overlooking Portland Harbor, the Combat Team moved to the Portland Docks on 9 March, loaded onto LCTs, and sailed on the afternoon of 10 March. At 0800 hours, the morning of 11 March, Exercise "Fox" began with the Battalion firing 105s from the LCTs, just prior to the landing of infantry on Slapton Sands. The guns unloaded on the beach from LCTs at 1300 hours and took up positions inland. The Battalion returned to Ivybridge at the conclusion of the problem, arriving early on the morning of 13 March.

The 29th Division now conducted an inspection of the Battalion, and General Gerhardt, Commanding General of the 29th, addressed the men, welcoming them to the division.

On 18 March, the Battalion moved to Braunton for training at the Assault Training Center. This particular course lasted until 30 March. Firing the M-7s from landing craft was practiced, and the Battalion made stout attempts to condition themselves on obstacle courses and cross-country marches. The obstacle course consisted of crawling under barbed wire, swinging hand-over-hand over wired pits, climbing cargo net, etc. This was intended to harden the personnel for the rigors of the invasion. The way the invasion turned out, more time should have been spent on swimming instruction and control of the bowels—but that is getting a bit ahead of things.

The return trip to Ivybridge was made on 30 March, and a period of service practice for the guns began. On or about 16 April, Headquarters Battery represented the Battalion in the four-mile speed march sponsored by the 29th. The march called for a full-field march in 45 minutes. The majority of the battery successfully completed it, but took days to recover fully.

On 20 April the Battalion again moved to the D-Camps at Dorchester, in preparation for another problem. There was some question in the minds of the men as to whether or not it was to be a problem or the real thing, but it turned out to be Problem "Fabius," a landing similar to the Problem "Fox." The 116th Combat Team had now been attached to the 1st Infantry Division for the invasion only.

In the middle of the night of 24 April, enemy planes attacked the docks at Portland. They also dropped bombs in the D-Camp areas, and bracketed one row of tents occupied by 58th personnel. Two men from Battery A were wounded, and a couple of tents flattened. On 1 May the Battalion loaded onto LCTs in Weymouth harbor, and on the morning of 4 May, landed on the beach at Strete. At the conclusion of the problem on 6 May, the Battalion departed for Bournemouth.

The week's stay in Bournemouth was very pleasant. The Battalion was quartered in buildings with real live bathtubs in them, of which good advantage was taken. It was also a week of springtime, and the sun outdid itself for us. For a while, it was almost possible to overlook the cloud of invasion hanging over us.

But on 9 May our Battalion CO, Lt. Colonel McQuade, received Field Order #35 from the 1st Infantry Division ordering the attack on France in the vicinity of Vierville-sur-Mer. At this time the plan was known only to a few Battalion staff members. Immediately the staff began a study of maps and aerial photos of the area across the English Channel.

On 15 May the Battalion departed for Area D-9, south of Dorchester via road and train. For the next few days, the 58th waterproofed equipment, gas masks, personal arms, etc. One of the memorable features was the amount of pork fed us. Pork chops, roast pork—the chow was good and there was plenty of it. We shook off that "Condemned men ate a hearty breakfast" feeling, and filled up.

On 29 May the entire Battalion was briefed on the coming operation and acquainted with the plans for landing on the coast of France. The Battalion was restricted to camp. No passes for any reason were issued after the briefing and the camp guard and restrictions were firmly enforced.

On 1 June, the Battalion split up, some remaining in D-Camps and the rest at the permanent camp at 45 Stourcliffe Avenue, Southbourne, England. The Reconnaissance parties, together with the Battalion CO's party, were to land with the infantry, and the remainder of the assault wave was scheduled to land on a progressive time schedule.

At this time the Battalion was assigned to the 1st U.S. Army, attached to V Corps, attached to the 29th Infantry Division, attached to the 116th Combat Team, which in turn was attached to the 1st Infantry Division for the forthcoming operation.

Loading on the invasion ships began on 1 June in compliance with Field Order #35, Headquarters 1st Infantry Division. The foot parties who were to go in first loaded on transports. The guns and a large portion of the Battalion personnel would disembark from LCTs, and a few from LSTs. We waited patiently in the harbor until 4 June, when the invasion fleet set sail. However, a few hours out, the fleet turned back to Portland Harbor as the invasion had been postponed 24 hours due to weather conditions.

In the early hours of 5 June, the invasion Armada got under way once more, sailed out into the Channel, and in almost parade ground formation started for the coast of France. As far as the eye could see stretched the ranks of LSTs, LCTs and APAs (troop ships).

D-DAY - 6 JUNE, 1944
OMAHA BEACH, FRANCE

H-Hour was at 0630 hours, 6 June 1944, and at that hour, under a cloudy sky and in rough surf, the first waves of the 116th Combat Team began unloading on Omaha Beach, near Vierville-sur-Mer, Normandy, France.

One FO, Lt. Carl M. Johnstone, and his party landed with the 5th Ranger Battalion at H-Hour, and attacked the cliffs at Pointe du Hoc. Three other foot parties, under Lt. Arthur Rosenbaum, Lt. Michael L. Runey, and Lt. Henry Shaddock, landed with the 5th Rangers at approximately H+40 minutes.

The remainder of the foot parties landed on Omaha Beach at 0700. They consisted of the CO's party, Liaison under Captain Alfred B. Fisher, Reconnaissance under Lt. Alfred E. DeMellier, Lt. Harry F. Russo and Lt. Samuel R. Kerr, and an FO party under Lt. Harry F. Grane.

Omaha Beach will become, unquestionably, another glorious page in American History. But on the morning of 6 June 1944, it was a cold, wet, shell-splattered hell, and it will be forever etched as such in the memory of the men of the 58th who participated in that landing. The beach and defenses beyond the beach had been thoroughly bombed and shelled by both warships and guns on the landing craft. But as the small landing craft grounded and their ramps went down, a murderous crossfire from well-emplaced machine gun nests cut them down. Mortars and artillery fire crashed along the water's edge, and men died in the water and they died on the shore. And all that long day, Death walked the beach at Vierville-sur-Mer.

Machine gun fire cut down the Battalion Commander, Lt. Colonel McQuade, and 1st Sgt. Hopkins as they struggled toward shore through waist-deep water. Lt. Russo was killed on the beach after assisting a badly wounded T/Sgt. Mason to shore. Lt. Grane was grievously wounded in the face, and two members of his party, Private Joslin and Private Jordet, were killed. Only one man out of this party walked off the beach.

The first waves were pinned to the beach by the withering enemy fire, and the Battalion's guns were unable to debark on schedule. But as the infantry gradually reorganized and worked painfully and slowly off the beach, the LCTs carrying the 58th's M-7s made it to shore through the incessant mortar, 88mm, and sniper fire. Some of the craft ran afoul of the mines at the water's edge, and one LCT sank with all the guns.

By 1800 hours, the Battalion had 11 of its guns on shore and ready to fire, and under the able and courageous direction of the Executive Officer, Major Walter J. Paton, moved up off the beach in close support of the infantry.

Looking back, it seems a miracle that anyone or anything ever got ashore. Casualties included 2 officers and 7 enlisted men killed and 3 officers and 14 enlisted men wounded and evacuated. In addition, 1 officer and 9 enlisted men were missing. Total losses of vehicles included five M-7s and trailers, 5 half-tracks, 4 quarter-ton trucks, 3 2 1/2 ton trucks, and 2 L-4 planes (the Piper Cub planes used for observation).

Personnel were scattered up and down more than a mile of beach, wet, shivering, and slightly bewildered. The beach was littered with wreckage and the still, twisted forms of men who had gone as far into France as they were ever going to go. And while at home the radio announced "operations are proceeding according to plan," and thousands sat in Soldier's Field and prayed in the afternoon sun, darkness settled on the bloody beach at Vierville-sur-Mer.

Nightfall saw the German night-raiders come out, to be greeted by a surprising display of ack-ack from the beach. Bombs were dropped along the beach and in the bay, but with little damage. Sniper activity continued throughout the night, but the shelling of the beach died down.

During the first hectic hours of the landing, the 58th provided the only effective artillery support on this stretch of the beach, and for the first few days, our FOs and Liaison provided the only communication for forward elements of the infantry and Rangers. The Battalion's FO party on Pointe du Hoc was isolated by the enemy for 2 1/2 days before the encircled Rangers were relieved by infantry and tanks of our forces. On D-Day, this same party, unable to get communication with the Battalion, had fired guns of the Navy on inland targets. The Navy continued to do excellent work on beach defenses and gun emplacements beyond the beach.

NORMANDY AND NORTHERN FRANCE

Our planes patrolled the skies on 7 June, and German air power was conspicuous by its absence. The 58th received first word that Lt. Colonel McQuade had been killed upon landing. Captain Wood and his section arrived on foot that morning. Their boat and all the equipment aboard it had sunk but all personnel were saved.

On the ground, the 58th continued to reorganize the Battalion while at the same time providing close support to the infantry (175th and 116th as well as the 5th and 2nd Rangers). Battery A was put in support of the 175th Infantry by Field Order #2. At 1800 hours B and C Batteries and the Fire Direction Center (FDC) moved to support the attack on Isigny-sur-mer. Sniping was bad now. The terrain inland off the beach was ideal for this kind of warfare, and snipers hung from the trees, from church steeples, and from the chimneys of buildings, and picked off our cautiously advancing troops.

On 8 June, A Battery rejoined the Battalion. We departed to a new position, still in support of the 116th Infantry, the 5th Rangers, and the 175th Infantry. We received two replacements of M-7s before moving to a new position northeast of Lison. Service Battery finally caught up

53

with us in this area, bringing more blankets, gas, food and water.

The fighting was now entering the hedgerow country of Normandy, excellent for defensive fighting, and the Germans were making the most of it. The Battalion was giving close support to the infantry, and was often only 1000 or 1,500 yards behind the front lines.

On 9 June the Battalion received orders to give direct support to the 175th Infantry, with one battery directed to support the 116h, which was still cleaning up the area from Maisy to Isigny. At 1445, Division suddenly realized the 116th needed more support and ordered us to give full support to the 116th and 5th Rangers while remaining in general support of the 175th. At 2130 the Battalion received 2 more M-7s, one to B Battery and one to C, for a total of 15 M-7s. At 2200 the Battalion departed for a new position.

Our Battalion was relieved from general support of the 175th and put in general support of the 115th Infantry on 10 June. The 115th hit some bad fighting on the river north of St. Clair. Our firing now was under the control of Colonel Cooper and the 110th Field Artillery Battalion.

On 12 June the Battalion fired over 1,800 rounds in rolling barrages with the intention of clearing the area across the river. (A rolling barrage is a technique where you fire a volley in front of your infantry, followed by a volley a short distance beyond the first, and a third volley a short distance beyond the second. This is supposed to clear the way for the infantry to move up in successive steps.) The mission was changed to general support of the 115th Infantry, and we went back in support of the 175th.

Later on 12 June the Battalion was relieved of attachment to the 29th Infantry Division and attached to the 30th Infantry Division, part of XIX Corps, in direct support of the 117th Division, who were just about ready to get their first taste of actual combat. One platoon of the 459th AAA (Anti-Aircraft Artillery) was attached to us at 2030 hours.

The Battalion CO, Major Paton, reported to General McClain, the Division Artillery CO, and the Battalion moved once again. Our mission was direct support of the 117th Infantry until it was found they weren't present. The mission then changed to supporting the 119th Infantry for an attack across the Vire River. One of our FOs was attached to the 119th to direct fire. The infantry attacked south between the Vire River and Taute Canal and gained the canal late at night.

On 16 June the Battalion moved to a new position near the town of Montmartin en Graignes. No sooner did we occupy the position than enemy small arms fire began sweeping the area. The following day enemy artillery fire fell in A Battery's area, wounding the Battery Commander, Captain Ralph E. Stone, and several enlisted men. That same day we received orders for

C Battery to join the 118th Field Artillery back near Airel. C Battery joined them at 1600 and returned to the 58th at 2230.

Our first mail arrived on 19 June—a brief but welcome distraction from the war. A slight change was made in position on 22 June, moving back about ten hundred yards. We maintained this position until 3 July, when the Battalion was attached to the 3rd Armored Division and moved to a position near La Madeleine.

On 27 June Cherbourg fell and everyone expected a big push—probably west of Carentan—but we remained where we were. On 29 June the Battalion fired its 10,000th round since the landing. In an irony of fate, the Battalion received orders for Lt. Colonel McQuade to report to Washington, D.C., PDQ. The order was dated 9 June. He died 6 June. At the end of June, the Battalion was assigned to the 1st U.S. Army, attached to XIX Corps, attached to the 30th Infantry Division, attached to the 120th Combat Team.

Following a meeting at Division Artillery to plan the attack across the Vire River and Taute Canal, the Battalion received orders effective 1 July attaching FO #2 to Combat Command B (CCB) of the 3rd Armored Division, detaching the Battalion from the 30th Infantry Division and attaching it to the 3rd AD. At noon on the 4th of July the Battalion, in conjunction with all other artillery on the front, fired one round toward St. Lo in salute.

The night of 6 July the Battalion moved

again, arriving in position near Airel. This was another hot spot when the Battalion first occupied it, with small arms and mortars kicking up quite a fuss. The 58th was up at 0300 and fired preparations for the attack at 0345. New orders attached us to the 120th Regimental Combat Team for an assault on St. Jean-de-Daye from the north. H Hour was at 1345, for which we fired a rolling barrage of 900 rounds.

On 10 July the Battalion moved near Cavigny, pulling into position shortly after midnight. It was fairly quiet when the position was first occupied, but shortly afterward it sounded like all the German artillery in that part of the country cut loose. The Battalion seemed to be about halfway between the artillery and their target. Reconnaissance went out at 1530 for new positions. We had a great shortage of M57 W.P. smoke, but the M84 base ejection was plentiful. The Battalion moved at 2200 to a new position. We were now supporting fires of the 391st Field Artillery Battalion.

A busy day ensued on 11 July. In the morning a young battle developed off to the left flank, and tanks in front threw a few rounds screaming over, but no casualties were sustained. The Battalion was assigned to the 1st US Army, attached to the 3rd Armored Division, attached to CCB, and in turn attached to the 30th Infantry Division. (In CCB we were second Battalion of the 391st Groupment and were under their control.) Patrols armed with bazoo-

kas were sent out. The 391st contacted two
tanks at 950 yards, and two more enemy
tanks were knocked out. The 58th fired
over 2,600 rounds.

At 0530 on 15 July the Battalion started
a rolling barrage at 0530 across the Vire
River. The 87th Armored Field Artillery
arrived to replace us in the 391st Groupment and the 58th moved to a position
west of St. Jean-de-Daye. One of our FOs
was wounded in the attack on the town
and he directed fire on the church steeple, which was being used as an enemy OP.
Several rounds later the steeple was gone.

On 17 July we drew 1800 rounds of Lot
No. ammo in preparation for a special shoot.
Our ammo allowance was reduced to zero
on 18 July. We now had 7000 rounds in
the Battalion. Service Battery was carrying
2100 in 9 2 1/2 ton trucks and trailers, and
firing battery was carrying 1370 on 6 M7s
with trailers.

On 19 July the Battalion was attached
to CCA of the 3rd Armored Division in
preparation for the forthcoming operation "Cobra." Plans had been perfected to
smash the defenses at St. Lo, and then to
break out of Normandy with an armored
drive. The operation was to follow an
intense aerial bombardment, after which
the armored penetration would proceed
south to Marigny then west to Coutances,
with the 3rd Armored Division spearhead-
ing the attack. All we were waiting for was
suitable weather for the bombers.

At about 0930 hours on the morning

of 25 July, the fighters showed up. P-47s swarmed all over the sky. Then came the Lightnings—diving and strafing the German ack-ack defenses. The big boys came parading in next, first the Liberators and then the Forts. After the Forts came the medium bombers.

The bombs could be seen falling like pepper from a giant shaker. Some fell short—especially on the 1st Battalion of the 9th Division. The ground started to shake, and it didn't stop for an hour-and-a-half. Ack-ack was intense at first, and two Liberators went down in flames. But the saturation of bombs eventually smothered all ack-ack. At that point the artillery cut loose. Somewhere along about 1100 hours that morning, the town of St. Lo ceased to exist.

Early in the morning of 27 July the Battalion took off, less C Battery, which had been attached to Task Force Y (TFY). The Battalion arrived in position at La Mesnil-Apry, where heavy enemy shelling was encountered by B Battery. In the early morning hours of 28 July, enemy planes bombed the area, injuring four men and damaging 3 half-tracks and a trailer. The Battalion moved again in the evening, and enemy planes again bombed the position, but inflicted no casualties or damage.

Travel on the road was becoming increasingly difficult, as the constantly moving columns jammed the roads day and night. As a result the Battalion was spending much of its time sitting on the road. Air cover was now complete in the daytime, and vehicles were now displaying cerise-colored panels so the planes could identify them as ours.

Early in the evening of 29 July, the Battalion occupied a position west of Cerisy-la-Salle. C Battery rejoined the Battalion, with FO Lt. Vincent Baker and his tank crew bringing 18 captured German prisoners to be sent to the rear. The 58th, now in general support of the 3rd Armored Division, moved to St. Denis le Gast on 30 July. The carnage was becoming horrific. A mixture of wrecked vehicles, dead Germans and dead horses lined the roads.

Having completed Cobra, and after taking over 5,000 prisoners, the Battalion moved again on 31 July, making halts along the road to permit supported troops to pass. That night the Luftwaffe really got in and pitched a fit. Virtually all their activity was at night now, and while their effectiveness was limited, they made things very unpleasant at times. Their routine was usually the same, as illustrated when the planes began milling around overhead. This continued for several minutes and then the Squadron Leader, or whoever he is, says, "This must be about the place, Hans."

So Hans proceeds to kick out about half-dozen or so flares, transforming night into day. They hang there, very prettily, drifting slowly to earth, with several thousand GIs cursing them every inch of the way. Just as they are about down, the overhead swarming seems to straighten out, and down they come for the bombing run.

This particular night must have been variety night on the Luftwaffe program, because one plane dropped H. E. bombs, another spilled incendiaries, and another must have had a lad in the rear dumping out small bombs from a basket. No one had any holes that night, as the Battalion had just pulled off the road. The area was pretty liberally plastered. Service Battery, some distance back of the Battalion, was getting its share of the treatment, resulting in 1 officer and 2 enlisted men being injured.

Losses in action July 1944

Sgt. Elmer R. Miller – 7 July 1944
Pvt. Leonard E. Needles – 11 July 1944
Cpl. Milton Gross – 27 July 1944
Pvt. Clayton D. Lohrmann – 27 July 1944
T/5 George J. Glennon – 28 July 1944
Pvt. Francis A. Biedrzyoki – 28 July 1944

On 1 August, the Battalion went into position north of Fouche in the evening. On 2 August the attack turned east at Brecey to outflank the German's left. We were on the right of the 1st Army, and the Third Army swung around us and headed for Paris. Lt. Shaddock received much fire in the L-4 Cub (plane), but was unhurt.

The following day the Battalion departed for a position at Juvigny Le Terte, but was held up on the road by anti-tank fire. One gun and one FO from A Battery were dispatched. The anti-tank gun was neutralized and the move to Juvigny completed. The Battalion remained at Juvigny until 5 August, firing intermittently and receiving slight counter-battery fire in A Battery's area that injured several enlisted men. Lt. Kerr finally located the enemy battery and we destroyed one gun of three. From Juvigny, a move was made to Ste. Marie du Bois, and from there to La Huvrie on the 6th. La Muzangeres was reached on 7 August.

During this march a 6x6 gas truck was destroyed by anti-tank fire on the road. The gas section had refueled the Battalion the day before and had to return to the gas dump many miles to the rear. After reloading they started out after the Battalion. The enemy had cut the road behind us, but the gas section didn't know this. So along in the afternoon here comes several 6x6s tooling merrily along the road when, BAM!, the leading truck went up in flames. The driver and assistant driver should have been cross-country runners judging from their stories of how they left the area.

On 8 August the Battalion was relieved from CCA and attached to CCB and attached to the 30th Infantry Division. The Battalion then moved back to a position in the vicinity of Juvigny Le Terte, where it remained, firing, until 13 August. During this period there was quite a bit of counter-battery fire throughout the area, but very little fell on the battery positions. Enemy planes were again very active, the more or less stabilized positions giving them opportunity to locate targets. On the night of 12 August, the area was bombed for 20 minutes, one bomb landing directly in B Battery's position. It caused no casualties, but did disrupt communications. Immediately thereafter an enemy artillery barrage came into the area, some falling in the CP area, but, again, causing no casualties.

On 13 August the Battalion was detached from the 3rd Armored Division to Corps Reserve (VII) for ordinance maintenance, and moved to a bivouac area near St. Mars de la Futaie. This was the first break of the campaign for the 58th. The bivouac area was on the edge of a small lake. Weather during the week stay there was ideal, and good advantage was taken of the opportunity to swim, bathe, and rest. Red Cross Club mobiles visited the area with donuts, entertainment and coffee. Several members of the Battalion took a side-trip to Mont St. Michel, the famous castle-cathedral on the coast, while others visited with neighboring French families. And those members of the Battalion of lower moral caliber also partook of certain amounts of Calvados—a particularly vicious native liquor.

At 0200 on 21 August, in a veritable cloud-burst, the Battalion was off to the wars again, in general support of the 3rd Armored Division. We coiled just east of Pre-en-Pail after traveling 67 miles. (Coiling is a maneuver of getting off the road still in column and making a huge coil in a field, sometimes a double or triple coil.) The Battalion followed the 991st FA (who had 155 mm howitzers, self-propelled) the next night for a march of 92 miles, arriving at an assembly area north of Courville at 0315. The Battalion moved again the next night and ended up 60 miles later just west of Corbeil.

The people were becoming increasingly glad to see the Americans as they pushed inland, and flowers and fruits in season were being cast into the vehicles as they rolled along. When the Battalion passed through Corbeil, tomatoes were in season. The Battalion crossed the muddy Seine River on a pontoon bridge about 25 miles below Paris on 27 August, and moved on to help Division mop up a little resistance at Lieusant. The Battalion left this position with B and C Batteries for a new position south of Brie Comte-Robert, on which we registered with high burst about 2200 hours from north of Reau.

When the column started moving the next morning, the forward elements met resistance, and one battery was placed in position to fire if needed. The resistance, two German tanks, was knocked out, and the Battalion moved on, stopping on objective near town of Montceaux-les-Provins.

The Battalion crossed the Marne River at La Ferte on 29 August and proceeded slowly north and east. The column was now passing through country familiar in American History—the Argonne, the

Marne, past Chateau Thierry, through Belleau Wood. In the space of 24 hours, the column rumbled past wheat fields and patches of woods where in 1917 and 1918 troops had slogged and crawled forward foot by foot. Occasional cemeteries containing the orderly rows of grey, weather beaten crosses of another war were passed, and in several towns and villages were the monuments raised to commemorate the American and British dead of 27 years past.

The Battalion got in position just east of Braisne at 0400 on 29 August. Here a long train of German supplies was caught and shelled. Tanks were left on the flat cars by fleeing German troops, as well as a nice load of perfume. We displaced forward several miles in order to attack over the Aisne River, but were not called on for fire.

At 1230 on 30 August the Battalion moved east, crossing the Aisne River at Bourg. The objective was to extend our beach-head line across the River. We proceeded six miles north to a small town named Pancy. At 1930 the Battalion started moving forward through forests, using leap-frog tactics, and finally took positions on the plain south of Laon, at 03-11. The Battalion was completely displaced at 2000.

The Battalion started a drive to the east on 31 August, and proceeded with little opposition through Bruzerer, Bestud, Sissonne, Lappion, Sevigny, Waleppe, Hanogne, and Seraincourt. At 1445 we met some resistance so B Battery went into position. By 1730 the resistance was cleared up and the

65

task force proceeded to our destination, arriving at 2130.

The Battalion's original objective was Nouvion, just south of Charleville, but after stopping just short of Herbigny, our mission changed from a drive eastward to a drive northward to Hirson. Some opposition was met just north of Rubigny, and the unit finally went into position there for the night.

That night the Battalion witnessed its first flying bombs. Quite a number of them came grinding over, and until they were finally identified they created quite a stir. They were apparently being launched from not too far away, as the "tail lights" could be seen rising in the distance.

Losses in action August 1944

T/4 Paul L. Hacker – 3 August 1944
T/4 Robert O. Fuller – 3 August 1944
Pvt. Orville A. Brenden – 3 August 1944
1st Lt. Alfred E. DeMellier – 9 August 1944

Awards presented August 1944

Date of Action

1st Lt. Samuel R. Kerr, Jr. – Silver Star
– 6 June 1944

S/Sgt Curtis Huffstickler – Silver Star
– 3 August 1944

T/5 Chester J. Kierakiewicz – Bronze Star
– 6 June 1944

T/5 Ralph J. Sullivan – Bronze Star
– 6 June 1944

Sgt. Joe N. DeWitt – Bronze Star
– 6-7 June 1944

Sgt. William R. O'Daniel – Bronze Star
– 6-7 June 1944

Pvt. Oscar Patrick – Bronze Star
– 6-7 June 1944

On 1 September, the column moved north to a position near Mt. St. Jean, and moved again that same day into position on the edge of the Bois d'Éparry, south of Hirson. Resistance was again met at the river crossing at Aubenton, and B Battery went into position at 1000 hours. By 1100 the entire Battalion was ready to fire.

But Colonel King, CO of Division Reserve, to which we were now attached, pulled away from this resistance and by passed it to the left. At 1500 we moved B Battery two miles west. At 1900 A bypassed B and stopped at Besmont. By 2000 the task force was crossing Than River at Bucilly and B and C Batteries moved with them to the objective, just short of Hirson. A Battery arrived at this position through the railroad underpass southeast of the town at 0145.

On 2 September Division attached again

with King's force on the right flank and moved in two columns north from Hirson. C Battery went with the right column, and the remainder of the Battalion with the left. Right column encountered considerable opposition a few miles out of Hirson. They finally extricated their force and by passed the enemy on the left route. The force moved slowly throughout the day and, at 2200, crossed the Belgian border.

BELGIUM

The Belgian people were much friendlier than the French. The column was stopped in the town of Grand Ring, where the men were overwhelmed with good beer, kisses and cries of welcome. The march continued until about midnight and, on 3 September, the Battalion occupied its first position south of Harverg.

This position was directly south of Mons, in the area of the one-sided battle of Mons in which German infantry and armor had been cut to pieces by our planes and advancing ground forces. As a result, isolated German units were wandering over the countryside at a loss as to what to do and where to go.

While the tail of the column was getting off the road, three German self-propelled 7 mm started through the area accompanied by infantry. One was knocked out, but one ran all the way through the bivouac until it bogged down in a ditch. Lt. Frawly captured five prisoners right across the road from our CP.

At 0630, another small tank battle broke out 300 yards west of our position. At this time CCA was northwest of us, just west of Mons, and CCB a few miles farther on. Upon arriving at Mons, the Division had trapped a large number of fleeing Germans.

We fired on their columns most of the day, and the P-47s worked them over again, setting many on fire. The first of 25,000-30,000 prisoners were captured.

The Battalion was now detached from the 3rd Armored Division Reserve, and attached to the 391st Groupment and CCB. The Battalion moved with CCB column on the morning of 4 September, 42 miles through Charleroi, Chatelet, and Fosse-la-Vie, on our way to a position at Floreffe near the Sambre River.

This march through Charleroi is memorable because of the hysterical crowds that lined the roads for approximately 25 miles, cheering and pelting the passing vehicles with blossoms, fruit, and bottles of beer. Passing through the city of Charleroi itself the column had difficulty negotiating through the packed throngs, and a halt almost meant disaster, because halted vehicles were immediately mobbed and the occupants hugged and kissed. Going through Charleroi the 58th looked more like a Tournament of Roses parade than a fighting unit on its way to battle.

American and English flags appeared from nowhere and fluttered from every building, along with the black, gold and red of the Belgian colors. American flags with all the way from 13 to 48 stars were seen, as well as red, white, and blue banners with the image of President Wilson—obviously holdovers from World War I. Cries of "Vive les Liberateurs" and "Allemand-Kaput" rent the air.

The Belgian Freedom Fighters (BFF) were very much in evidence, some of them sporting a sort of linen beach suit uniform, with black tams. They buzzed gallantly around the towns, armed to the teeth, piled on automobiles of dubious mechanical performance, much to the delight of the homefolk. They were engaged in rounding up collaborationists and other questionable characters, and patrolling the side-roads. They made it very difficult for the scattered German troops to hide out successfully for very long.

The Battalion stopped as planned just short of Namur, on the bend of the Meuse River. As we drove into a farmyard, the owners were taking off thinking we were Germans. They were overjoyed to find us Americans and offered us everything in sight. We made our CP in their shed and coiled for the night despite civilian reports of Germans in the woods.

At Floeffe, early on the morning of 5 September, enemy mortar and artillery fire began falling in the Battalion area. Observers spotted enemy installations across the valley, and direct fire from 105s and AA was brought to bear on them. Much of the fire was directed by an English speaking Baron, who telephoned across the canal and got reports on the Germans from his friends. One of those friends reported two machine guns on either side of his house. With this help, we drove them across the ridge. Ten casualties were suffered by the Battalion, and three of them had to be evacuated. At 1600 the Battalion moved a few miles southeast and put our CP in another barn.

On the morning of 6 September, the Battalion passed through more cheering throngs in Namur, crossed the Meuse River with the Combat Command, with the mission of supporting the crossing of the 9th Infantry Division troops further down the river. The guns went into position off the road at one point when an enemy column was observed, but effective fire could not be brought on it. Instead, we fired on enemy infantry from a position in a field near the river crossing of the 9th Infantry, using Charge 1. From a position behind the guns the projectiles could be seen in flight, arcing into the sky like a hard-hit fly ball headed for deep field. At 2000 the Battalion left on a hectic night march in the driving rain, traveling down a narrow gorge that was darker than the inside of a black cat, and finally coiling in an area on the Meuse River near Dinant.

Prisoners were brought in the next morning, herded in by a half-track, which struck a mine while passing through the Battalion position, injuring two of the crew. The column moved shortly thereafter and halted on the road at Dorrine, while C Battery fired on enemy infantry. The Battalion then moved on to Sovet, where it coiled to repair vehicles before moving again in the late afternoon to rejoin the 3rd Armored Division near Leige. We bedded down for the night on a farm just short of Villersle-Tempte, where a woman who spoke English fixed us up with wine from the local Priest, fresh eggs, and even milk.

The Battalion met up with the 3rd Armored Division in the vicinity of Tilff, where CCB was already in position. Between 1400 and 1600 we moved forward to a firing position just east of Romsie. This night Lt. Hipp and three other men captured 145 German prisoners and held them all night with the help of the BFF.

On 9 September, CCB was given the mission of occupying the high ground to the southeast of Liege. We proceeded slowly because of enemy tanks and infantry, and left the rest of the Battalion in position to cover our advance. P-47s bombed and strafed ahead of us. Once CCB was in position, the huge slag mountains from the smelters of Liege were visible, each with the colors of Belgium flying from their peaks.

At 1320 on the 9th we received word that our CO, Lt. Colonel Paton, had been wounded in the right arm by a smoke shell (M84) fragment while adjusting B Battery fire on an enemy target. He was evacuated, and Major Starling W. Wood, Jr. then assumed command of the Battalion. C Battery went into position southwest of Bannage, and our planes could be observed steadily and relentlessly strafing an enemy column northeast of the town. At 1745 our P-47s had to leave the tanks alone and shoot down two enemy planes. One landed a couple hundred yards from the CP, and the pilot was captured. At 2030 we tried to proceed forward but were stopped after a mile as the bridges were blown up ahead.

The morning of 10 September the Battalion moved to a position southwest of Verviers. Colonel Gaston, CO of the 391st, was also wounded, so the 58th was detached from CCB and attached to CCA. From 1700 to 1845 we moved north of Verviers, overlooking the town. Verviers was also very enthusiastic about the Americans. But from the hill above the town we could see the black rolling hills of Germany. It was questionable if the reception there would be quite as delirious.

On 11 September, the Battalion was attached to TFX. At 1040 we moved forward but by 1230 were halted by enemy tanks. Battery C led off with TFX on the morning of 12 September, with the rest of the Battalion following, and at 1632 our leading elements crossed the German border.

GERMANY

The Battalion remained in position on German soil on 13 September, firing on pill boxes. Ammunition became very critical at this point, and the 58th borrowed 1,000 rounds from the 67th Field Artillery. On 14 September we passed through the dragon's teeth, breaching the vaunted Siegfried Line, and went into position one mile north of Brand. On 17 September, one of our FO tanks was knocked out, injuring two of the crew.

The Spearhead drive had come to a halt, not to be resumed for many months. The Battalion remained in position at Brand until 21 September, firing steadily in support of the 3rd Armored Division and 1st Infantry Division who were rammed against the town of Stolberg.

On 21 September, the Battalion finally moved to a position at Busbach, Germany, the guns going into positions directly in back of the town. The Commanding General of the 3rd Armored Division, Major General Maurice Rose, visited the Battalion and presented awards and promotions to individual officers and enlisted men of the Battalion. We remained in position at Busbach, firing consistently on targets in Stolberg and surrounding areas.

From 22 September to 30 September,

the Division assumed a defensive position while the 1st Infantry Division and the 9th Infantry Division consolidated their positions. The 3rd Armored Division rotated the units on the line to give all a rest and maintenance period.

Losses in action September 1944

Pvt. Eddie Valdez – 5 September 1944
S/Sgt. Roy J. Walters – 5 September 1944
Sgt. Rosebur J. Miller – 5 September 1944
T/4 Robert Peiser – 8 September 1944
Major Walter J. Paton – 9 September 1944
Pvt. 1st Class Frank L. Lapsley – 15 September 1944
Sgt. Otto G. Hultgren – 15 September 1944
S/Sgt. Jack James – 16 September 1944
T/5 Laudie P. Jilek – 18 September 1944
Cpl. Claude Z. Watt – 26 September 1944
Pfc. Kenten R. Enbody – 26 September 1944
Pfc Otis J. W. Brown – 29 September 1944

Awards presented September 1944

Date of Action

Captain Paul K. Switzer – Silver Star – 6 June 1944

T/Sgt Joseph V. Rosko – Bronze Star – 15 July 1955

Sgt. Charles P. Hunt – Soldier's Medal
– 14 August 1944

T/4 Harley L. Rountree – Bronze Star
– 6 June 1944

T/5 Robert F. Beinlich – Bronze Star
– 17 June 1944

T/5 John C. Blum, Jr. – Bronze Star
– 17 June 1944

T/4 Joseph F. Jana – Bronze Star
– 28 July 1944

Cpl. Alvin Freeman – Bronze Star
– 17 June 1944

Pfc. Truman W. Gee, Jr. – Bronze Star
– 17 June 1944

1st Sgt. Elwood J. Monighan – Bronze Star
– 6 June-8 Sept. 1944

Pfc. Joseph Priester – Bronze Star
–17 June 1944

1st Lt. Henry Shaddock – Air Medal
– 29 June-28 July 1944

1st Lt. Leroy O. Stevens – Cluster to Air Medal
– 22 June-10 July 1944

2nd Lt. Charles R. Snyder – Cluster to Air Medal
– 22 June-13 July 1944

S/Sgt Thomas K. Turner – Cluster to Air Medal
– 23 June-14 July 1944

Service

S/Sgt Jack James – Bronze Star –
6 June-10 September 1944

S/Sgt Roy J. Walters – Bronze Star
– 6 June-10 September, 1944

Sgt. Otto G. Hultgren – Bronze Star –
June-10 September 1944

T/4 Henry R. Borter – Bronze Star
– 19 March 1943-10 Sept. 1944

T/4 Robert E. Lee – Bronze Star
– 19 March 1943-18 Sept. 1944

Pfc Frank Flores, Jr. – Bronze Star
– 19 March 1943-18 Sept. 1944

On 1 October, the Battalion was detached from the 3rd Armored Division and attached to the 1st Infantry Division for the attack on Aachen. The 58th moved to a position seven miles southwest of Aachen in support of the 1106th Engineer Groupment, consisting of the 237th Engineer Battalion and the 238th Engineer Battalion. There we fired only prepared concentrations in support of the upcoming attack.

The plan now was for XIX Corps to attack from the west, bypass Aachen on the north side and secure the high ground northeast of the city. When this was accomplished the 1st Infantry Division was to attack from the south and join XIX Corps in cutting off the city. The 1106th Engineers were to make a demonstration and feint attack with the 1st Division attack. The 58th was to support the Engineers in this demonstration.

Lt. Colonel Mont Hubbard was attached to the 58th as Battalion Commander on 2 October. Captain Smith was promoted to Major and Lt. Frawley to Captain, effective 29 September.

The Battalion remained in position, firing the prepared concentrations until 8 October, when we fired 20 rounds of propaganda leaflets into Aachen. On 9 October the Engineers loaded a trolley car with TNT and rolled it down the hill toward the Germans. It made a big noise.

The 1st Infantry Division sent an envoy to Aachen with surrender terms on 10 October. The envoy was blind folded and taken to German headquarters with the message that the Germans must answer by 1030 on 11 October. The Germans refused to surrender, and our artillery and air force both opened up on Aachen. Visibility was good, and on 12 October the artillery alone poured 169 tons of explosives into Aachen. The Battalion marked several targets with red smoke for the fighter-bombers.

Aachen was now almost completely surrounded. XIX Corps was coming down from the northwest and the first Infantry Division was coming up from the southeast while the 58th continued to pound the city with artillery. During this operation, the Battalion had one FO in Holland, a second in Belgium, and a third—along with firing and Service batteries—in Germany.

The Service Battery had 155 Howitzers in back of it. These guns attracted considerable attention from enemy artillery and planes, with the result that life in Service Battery was not particularly dull for a few nights. The rainy season was in full swing and the first barrage to come over caught the battery with slit trenches full of water. A couple nights later the Germans air-raided the area five times in one night, employing a unique technique of dropping the bomb first, then hanging out flares and strafing. Service Battery soon moved over with the Battalion, but the Air Section stuck it out until a direct hit on a chicken coop in the back of the building in which they stayed gave them march order.

Lt. Alfred DeMellier returned for duty on 12 October. FO Vincent Baker was promoted from 2nd Lieutenant to 1st, effective 5 October. The Battalion supported the 1106th Engineer Combat Group in retaking Hill 203, an action that resulted in a

Commendation for the Battalion. (See in its entirety at the end of October.)

An enemy counter attack on the 13th was broken up by an excellent combination of infantry and artillery. The Battalion again marked targets for dive-bombers in Aachen, and the bombers worked it over so steadily that a yellow cloud hung over the besieged city for hours at a time.

The encirclement of Aachen was completed by 17 October. On 20 October, the 58th fired 50 more rounds of leaflets. Finally, on 21 October, we received word that the city had fallen. On the following day, 22 October, we were detached from 1st Division and reattached to the 3rd Armored Division. Back went the 58th to its former position in Busbach. At this point a system was instituted whereby the Battalion was to be five days in direct support and five days in reserve.

The Battalion took over from the 67th Armored Field Artillery Battalion at 1200 hours on 25 October, and at 1202 was given a T.O.T. We had two observers in Stolberg and one at a rear OP Our sector was only 1,500 yards wide, with the 1st Division on our left and CCB on our right. The Battalion was also to reinforce any fires needed in the adjacent sectors.

Seven men were wounded by enemy fire on 28 October, while five were wounded on the 29th. The medical 1/4-ton was damaged when sent to aid our wounded men. Also, a 1 1/2-ton maintenance truck was demolished when hit by enemy shell fire.

On 30 October the enemy fired one of the heaviest barrages that the Battalion had encountered since landing in Europe. Over 150 medium caliber shells landed just beyond the Battalion position near the 155 S.P. Gun Battery. The adjustment started with seven airbursts fired over A and B Batteries, wounding one man. But in the fire for effect the range was increased seven hundred yards and no other damage was sustained by this barrage. On 31 October we registered the Battalion, using the 67th Armored Field Artillery Battalion observer. Major Bradford E. Smith left the 58th and was assigned to the 406th Artillery Group as S-4 (Supply Officer).

Losses in action October 1944

T/5 Gilmer L. Cowan
– 25 October 1944

Pvt. Luther B. Petty
– 26 October 1944

Pvt. David R. McAllister
– 27 October 1944

Pvt. Morgan VanLandingham
– 27 October 1944

Capt. Edward L. Chadick
– 27 October 1944

Pfc. Joseph S. Tackett
– 27 October 1944

Pvt. Earnest P. Howard
– 27 October 1944

Awards presented October 1944

Date of Act

Pfc Frank Flores, Jr. – Silver Star
– 8 September 1944

Pvt. Arthur W. Frost – Silver Star
– 8 September 1944

T/5 Marshall E. Olsen – Bronze Star
– 29 September 1944

Pfc Thomas L. Brooks – Bronze Star
– 29 September 1944

1st Lt. Leroy C. Stevens, Jr. – 2nd Cluster
to Air Medal
–13 July – 3 August 1944

2nd Lt. Charles R. Snyder – 2nd Cluster
to Air Medal
–13 July – 10 August 1944

S/Sgt Thomas K. Turner – 2nd Cluster to
Air Medal
–14 July – 9 August 1944

T/4 Harold S. DeHart – Soldier's Medal –
–29 September 1944

T/4 Robert R. Wright – Soldier's Medal –
–29 September 1944

T/5 Delinal Russell – Bronze Star
– 15 September 1944

T/5 Virdi M. Larson – Bronze Star
– 15 September 1944

1st Lt. Henry Shaddock – Cluster to Air Medal
–29 July – 15 September 1944

T/5 Reuben E. Rudd – Silver Star
– 2 September 1944

Cpl Stephen Kohan – Bronze Star
– 15 September 1944

S/Sgt Jack James – Cluster to Bronze Star
– 3 September 1944

S/Sgt Louis P. Righero – Bronze Star
– 6 June 1944

Capt. Anthony B. Montana – Bronze Star
– 11 August 1943

1st Lt. Harry F. Grane – Bronze Star
– 11 August 1943

Capt. Ralph E. Stone – Bronze Star
– 6 June 1944

1st Lt. Carl M. Johnstone – Bronze Star
– 12 June 1944

Service

Capt. Harold E. White – Bronze Star
– 2 Nov. 1943 – 6 June 1944

1st Sgt Clarence W. Durham – Bronze Star
– 2 Nov 1942 - 18 Oct 1944

T/Sgt Wilton May – Bronze Star
– 2 Nov 1942 - 18 Oct 1944

Sgt Anthony P. Bracaglia – Bronze Star
– 2 Nov 1942 - 18 Oct 1944

T/4 Ralph J. Raimo – Bronze Star
– 2 Nov 1942 - 15 Oct 1944

T/5 Gerald L. Darby – Bronze Star
– 11 Nov 1942 – 13 Oct 1944

S/Sgt George J. Storer – Bronze Star
– 2 Nov 1942 – 15 Oct 1944

T/3 John R. Williams – Bronze Star
– 2 Nov 1942 – 8 Oct 1944

Sgt. James P. Lage – Bronze Star
– 2 Nov 1942 – 15 Oct 1944

Cpl Wilbur E. Joslin – Bronze Star
– 2 Nov 1942 – 1 Oct 1944

FRAN BAKER

Capt. George L. Hagerty – Bronze Star
– 2 Nov 1942 – 1 Oct 1944

CWO John T. Eastburn – Bronze Star
– 2 Nov 1942 – Oct 1944

S/Sgt Horace G. Ward – Bronze Star
– 14 June 1943 – 8 Sept. 1944

Commendation

Subject: Commendation, 58th Armored Field Artillery Battalion
To: Commanding Officer, 58th Armored Field Artillery Battalion
Thru: Commanding General, VII Corps, APO 307, U. S. Army

From Headquarters, 1106th Engineer Combat Group
Date: 24 October 1944

1. Beginning September 28th and continuing until October 21st the 1106th Engineer Combat Group, then attached to the First Infantry Division, was directly supported by the 58th Armored Field Artillery Battalion in the defense of a sector South and Southwest of Aachen, Germany. Without this support our mission could not have been accomplished.
2. One outstanding feature always evidenced in the performance of the 58th Armored Field Artillery Battalion was the spirit of true cooperation on the part of

86

every individual of that organization. This cooperation meant all the more to The Engineer Group in view of the fact that this was the first experience of the Group as a whole acting on an infantry mission. Battalion commanders, company commanders, platoon leaders, and staff officers of the 1106th Engineer Combat Group are all unanimous in their praise of the support rendered by the 58th Armored Field Artillery Battalion.

3. The performance of First Lieutenant Vincent Edward Baker is particularly deserving of mention. As Forward Observer, he volunteered to accompany an engineer platoon in the capture of Hill 203, a critical terrain feature previously captured by our forces but from which we were forced to withdraw. Lieutenant Baker accompanied this platoon on 12 October and while the engineers were consolidating their positions brought down on the enemy such an accurate artillery barrage that the enemy could not counter-attack. Lieutenant Baker then established a system of defensive fire which enabled the engineer platoon to hold the position against superior enemy forces.

4. It is with sincere gratitude that I take this opportunity to commend a very fine organization.

/s/ T. DeF. Rogers
Colonel, OE, Commanding

On the night of 1 November, the first heavy caliber shells, probably 170 mm, were pumped in steadily for approximately 45 minutes. A total of about 150 rounds landed just over the Battalion position, presumably directed at the 155 guns in back of the Battalion. Basements and foxholes became extremely popular for a while, but no damage or casualties resulted in the Battalion.

The rainy season had definitely set in now, and things were pretty sloppy. The Battalion was still in groupment with the 67th Armored Field Artillery Battalion in support of CCA and the 3rd Armored Division was still in a defensive position. We held the high ground southeast of Stolberg on a line to Mausbach as we continued alternating direct support with the 67th.

The Battalion received verbal orders on 3 November that we were detached from CCA and attached to CCB in a groupment with the 391st Armored Field Artillery Battalion. Their sector was the right half of the Division's sector. Because of the danger of counter attacks it was decided that all OPs would be manned jointly by both the 391st and the 58th in order that an observer could be on duty 24 hours a day. The mission remained the same—to defend the high ground southeast of Stolberg.

Enemy activity was more or less limited to the nights. The Luftwaffe did make an occasional appearance during daylight hours when they could get "road clearance" from the AAF. The Battalion was fortunate in having attached to it a very excellent ack-ack outfit, the 486th, with multiple .50 Caliber guns and 37mms. These boys were with the Battalion all during its attachment to the 3rd Armored Division, and became more or less a part of it instead of just an attachment. Their multiple .50s supplied extremely effective firepower, not only against planes, but also against ground troops.

On 4 November the Battalion sent one Liaison Officer to Task Force Lovelady, and three observers to the OPs manned by the 391st. It was also decided that we would register in both sectors each day in order that we could reinforce the fires of the 67th Armored Field Artillery Battalion. On 6 November the Battalion fired several precision adjustments on an enemy OP and scored many hits on each mission. Using Fuse T105, we had a much better effect on the building than we would have had using fuse delay.

The Battalion continued to receive enemy shelling, but sustained no damage or casualties. There was an increase in enemy activity along the front in our sector. One enemy patrol was spotted by our observer on the left front of our sector, and several Battalion volleys dispersed this patrol. We now left our guns laid so that they boxed in each observer or the observer whose OP was in the "hottest" sector.

The Battalion received instructions on

10 November about coming operations. All observers were recalled, but Liaison remained with Task Force Lovelady. Four FOs and one other Liaison Officer were sent to Task Force Mills, also one FO and one Liaison Officer to CCB Reserve. The objective of Task Force Mills was the towns of Hastenrath and Scherpenseel. Task Force Lovelady's objective was in the vicinity of Werth. With Task Force Mills we had one observer with each tank company and one with the infantry company.

Fire plans for the coming attacks were given to the 58th's observers on 11 November, and final plans were made for the attack. A little snow fell in the 58th's area on 13 November, decreasing visibility as enemy shelling continued to increase. Approximately 80 rounds of 106mm fell in and just over the Battalion's area on 14 November. There was no damage, but two casualties were sustained by the 58th. A battery of assault guns in position behind us had 10 or 12 casualties.

D-Day was 16 November, and H-Hour was set for 1245. Visibility was better, and our planes were overhead all morning, bombing and strafing enemy positions. Battalion artillery preparations started at 1115, and we fired 760 rounds up to H-Hour. The attack immediately ran into difficulty. Our tanks bogged down, and this, coupled with stubborn defenses, slowed the attack to a walk. Most of the missions fired were unobserved or by Air OP. Four of the Battalion's FOs – 2nd Lt. John R. Jackson, Jr.,

2nd Lt. James F. Cragin, 2nd Lt. James C. Hipp, and 1st Lt. Henry Shaddock – were wounded and evacuated. All three of the Battalion's FO tanks were knocked out of action. The forward elements of Task Force Mills did not reach their objective and stopped just north of Werth for the night. The Battalion boxed in the forward positions and fired harassing and defensive fire around the tanks all night.

What was left of Task Force Mills joined forces at Werth on 17 November, and at 0800 started for their objective. Colonel Mills, the 58th's observer, and five medium tanks reached the left objective, Hastenrath, at 1000 hours. Another of the Battalion's observers, accompanying a small group of infantrymen, reached the right objective, Scherpenseel. The Battalion's observer at the left objective had been wounded by sniper fire, so another officer was sent to take his place but couldn't reach the objective on foot during the daylight hours. The right objective was counter-attacked for two hours in late afternoon, but the Battalion broke up the attack using time fire on the German infantrymen.

On 20 November the task force supported by the Battalion was "pinched out" of the action by the 1st Infantry Division and the 104th Infantry Division coming up and closing in from the flanks. All our observers who hadn't been wounded were recalled, but Liaison was kept with the unit commanders at both objectives. The Battalion sent two more officers to the hospital, giving us a total strength of 24 officers. We fired a smoke mission for the 1st Infantry Division in preparation for an attack.

CCB was now placed in reserve, so the 58th was attached to CCA under General Hickey, in a groupment with the 67th Armored Field Artillery Battalion. We were in direct support of TFY under Colonel Doane. On 26 November the Battalion moved to the vicinity of Eschweiller to support the task force in the attack on Langerwehe, but returned the next day to the old area at Busbach and was placed on a two-hour alert status. Lt. Colonel Paton returned to the Battalion from the hospital on 30 November.

Losses in action November 1944

PFC Edwin M. Stephens
– 3 November 1944

Sgt. Forrest L. Hartzell
– 7 November 1944

T/4 Bayne J. Contz
– 7 November 1944

2nd Lt. John R. Jackson, Jr.
– 16 November 1944

PFC Steve R. Hanson
– 17 November 1944

2nd Lt. James C. Hipp
– 19 November 1944

2nd Lt. James F. Cragin
– 19 November 1944

1st Lt. Henry Shaddock
– 19 November 1944

1st Lt. Vincent E. Baker
– 20 November 1944

Sgt. Jack H. Boykin
– 23 November 1944

Pvt. James C. Coleman
– 24 November 1944

Awards presented November 1944

Date of Action

T/4 John H. Glass – Bronze Star
– 11 July 1944

Cpl Manfield L. Miller – Bronze Star
– 11 July 1944

Pvt Arnold L. Kopecky – Bronze Star
– 11 July 1944

Service

Major Bradford E. Smith – Bronze Star
– 2 Nov. 1942 – 6 June 1944

Capt. Lewis B. Fisher — Bronze Star — 2 Nov. 1942 — 6 June 1944

On 1 December the Battalion moved to a position south of Heistern to reinforce fires of the 391st Field Artillery Battalion. All batteries were registered by Air Observer in spite of poor visibility. The attack was postponed for 24 hours and then called off, so the Battalion returned to Busbach.

Lt. Colonel Hubbard was relieved of command on 4 December, and Lt. Colonel Paton resumed command of the 58th.

On 9 December the Battalion departed from Busbach and again took positions in the vicinity of Heistern. There we were to be in general support of the 391st, to support Task Force King of the 3rd Armored Division. This task force was to attack from Langerwehe in an attempt to secure the towns of Obergeich, Geich, and Echtz. The attack was to be coordinated with the attack of the 9th Infantry Division to secure the west bank of the Roer River near Duren. The Battalion was to place fire in front of the advancing tanks to prevent the enemy from using his infantry bazooka teams, which had been knocking out our tanks of late.

The prepared fires were laid down on the morning of 10 December, and the attack took place on schedule. The Battalion continued to fire harassing fires throughout the day. The attack went well, and Geich was reached on schedule. Echtz was taken the following day. Task Force King reached the river at several points, and the enemy was pushed out of Hoven and Mariaweiler. The Battalion fired on AT guns and groups of enemy infantry with good effect.

A smoke mission was fired across the Roer River on 12 December—the Battalion's only mission for the day. All objectives were secured, and the enemy had retreated across the river. On 13 December the Battalion moved back to the area of Busbach.

The evening of 16 December orders were received relieving the Battalion of attachment to VII Corps and attaching it to the 28th Infantry Division and VIII Corps. The Battalion was to proceed to Wiltz, Luxembourg, to reinforce fires of the 28th, which was being counterattacked. It was rumored around the Battalion that they were "having a little trouble"—which is probably the most colossal understatement ever made.

THE BATTLE OF THE BULGE - BELGIUM

During the night of 16 December enemy parachutists were dropped near the Battalion, but they were rounded up and captured. The Battalion Commander departed on the morning of the 17th to contact the 28th Division Artillery Commander. Road clearance for the 58th could not be obtained, so movement was postponed until 0200 hours on 18 December.

The Battalion departed from Busbach, Germany, enroute to Bastogne, Belgium. It was a gala night for the Luftwaffe. There were enemy planes overhead almost constantly, dropping flares and bombs near us, but we suffered no damage. The flares made it easier to see the road, but made it hard as hell on the nerves. No casualties resulted, however. The Battalion passed through Verviers about 0800 hours on the morning of the 18th.

As we approached our destination the weather became more overcast, and fog began closing in. Also there seemed to be a growing agitation among the civilians we passed on the road. When we passed through Houffalize the civilians were streaming out of town carrying their belongings. What's more, they looked scared. The question began to arise: "What in hell is going on down there?"

When the column passed through Bastogne people were milling around all over the place—and they were distinctly worried. Service Battery and Personnel Section remained in Bastogne. The rest of the Battalion moved on out of town, past units who were digging in alongside the road, and took up defensive positions near Longvilly, Belgium, near the Luxembourg border, to support fires of the 100th Field Artillery Battalion, which was in direct support of the 110th Infantry Regiment, 28th Infantry Division.

At 2200 hours small groups of our infantry began infiltrating back through the Battalion position with information that a strong enemy force was advancing down the Bastogne road. The Battalion Commander, Lt. Colonel Paton, elected to remain in position and attempt to stop the enemy armor and inflict as much damage as possible. Personnel of the Battalion were informed of the situation. Positions were dug in and out-posts posted.

The Battalion organized four M-10 tank destroyers that were pulling back through the position, and together with the infantry, placed them in front of our position and on our flanks. Throughout the night of 18 December fire was directed on the main road crossing, where enemy troops were concentrated. In the early hours of the 19th, heavy mortar fire began falling in the Battalion position. Two enemy half-tracks attempted to move up along the road toward A Battery's position, were taken under direct fire, and knocked out.

A heavy fog made observation impossible. Soon heavy small arms fire began sweeping the area from the front and both flanks. The infantry began to withdraw, and it quickly became apparent that the 58th held the forward-most position in the sector.

With improved visibility, two enemy tanks were located on the hill above the position, throwing direct fire into the batteries. Taken under direct fire by our guns, they were forced to withdraw. The Commander of TFR of the 9th Armored Division decided to move his forces back to Bastogne. With no infantry or tank support, the 58th was also forced to withdraw to avoid being overrun by the enemy.

Near Mageret the head of the column ran into a road block, and attempts to break through were unsuccessful. B Battery remained in position at Longvilly to cover the withdrawal of the column and fired direct fire on enemy tanks throughout most of the morning. The other two batteries went into position to fire on Mageret, but limited observation made effective fire difficult.

The enemy continued to shell the road, and mortar and machine gun fire swept most of the column on the road. Wrecked and burning vehicles jammed the road. B Battery was catching direct fire from enemy tanks, inflicting losses in vehicles, guns and personnel. Passage through Magaret being impossible, the column bypassed the town on a road to the right of the town. Movement was possible, though enemy fire continued to fall on the road.

The Battalion had suffered casualties to the extent of two officers killed, 29 enlisted men killed or missing, and 58 wounded. Other losses included 8 M-7s, two M-4 FO tanks, 14 half-tracks, 2 3/4 ton trucks, 1 3/4 ton ambulance, 1 wrecker, and 10 1/4 tons.

The remainder of the Battalion went into position at a point just west of Bastogne in the afternoon, in support of the 101st Airborne Infantry Division. The delaying action fought by the Battalion that morning had enabled two task forces to take defensive positions behind them, and measurably slowed the enemy advance on Bastogne. A number of missions were fired by the remaining 8 guns of the Battalion on the 20th, covering road blocks held by the 101st. On one road block Battalion fire destroyed 9 enemy vehicles attempting to break through.

On the morning of 21 December, word was received that the enemy had broken through on the right flank. The Battalion moved to a position near Tillet to more effectively cover the road blocks, arriving in position at 1300 hours. Soon thereafter the position was surrounded, and all roads leading out of Tillet were cut. Two M-7s and an M-4 tank, on reconnaissance to find a way out, were destroyed by fire from an enemy road block. The Battalion set up outposts and road blocks for the night.

Shortly after midnight the Battalion received orders to attempt to return to Bastogne. A small force of friendly armor had been sent out to meet us. The Battalion moved out on the road in column, and after moving about 500 yards ran directly into an enemy column advancing down the road toward the Battalion position.

Our lead M-7 was knocked out by AT fire, and the column engaged in a fire fight, receiving machine gun and mortar fire. The original position was resumed, and the remaining 5 M-7s brought fire on a strong enemy force moving up to attack in the morning. By 0930 hours on 22 December all but one of the Battalion's M-7s had been destroyed, and the order was given to destroy all remaining equipment as it was impossible to hold out any longer and to infiltrate through the enemy lines on foot.

The men were sent out in small groups through a heavily wooded area which had not yet been occupied by the enemy, leaving the remains of the Battalion blazing in the snow. A heavy snowstorm aided the withdrawal. The men walked about 17 miles that day, before friendly units were contacted. All who left the position had gotten through safely.

Service Battery had withdrawn from Bastogne on the morning of the 19th and arrived at our rear echelon located at Fontenoille, Belgium on the 21st. Late at night on the 22nd the first survivors of the Battalion reached the small hamlet, and the remainder continued to come in for the next few days. Some were taken back to Charleroi before locating the remainder of the Battalion. Thus, the 58th Armored Field Artillery Battalion, less Armor and less Field Artillery, assembled at Fontenoille to lick its wounds, count its losses—and sweat out transfer to the infantry.

The people of Fontenoille took the 58th into their homes, and for three weeks it was just like being "one of the family." The heartwarming hospitality they exhibited toward the rather battered Battalion overcame the barnyard stench that permeated each and every dwelling. The 58th will ever have a place in its heart for the Belgian people of Fontenoille.

The Battalion was attached to the 174th Field Artillery Group on 24 December, and operated road blocks during this period as a precaution against further penetration. A small party of the Battalion that had gone into Bastogne after the battle of Longvilly returned to the Battalion on the 28th after that city (Bastogne) was relieved. Lt. Baker returned to duty, along with a number of replacements for the men we had lost. On the 29th of December the Battalion was relieved of attachment to the 174th and was attached to the 333rd Field Artillery Group.

New equipment began to arrive on 26 December, and we started whitewashing it to blend in with the snow-covered landscape.

101

Losses in action December 1944

Cpl. Stephen Kohan
— 6 December 1944

Pvt. Charles L. Roberts
— 18 December 1944

Captain Anthony B. Montana
— 19 December 1944

1st Lt. Alfred E. DeMellier
— 19 December 1944

Tech Sgt. Joseph V. Rosko
- 19 December 1944

S/Sgt Elmer R. Hottell
- 19 December 1944

S/Sgt Melvin W. Padgett
- 19 December 1944

S/Sgt James E. Watkins
- 19 December 1944

Tech 3 Ralph J. Raimo
- 19 December 1944

Sgt. Walter J. Abramski
- 19 December 1944

Sgt. James L. Caddell
- 19 December 1944

Sgt. Evan B. Lum
- 19 December 1944

Sgt. Joseph T. Thompson
- 19 December 1944

Sgt. Harmon F. Yates
- 19 December 1944

Tech 4 Owen D. Marshall
- 19 December 1944

Cpl. August T. Klopsch
— 19 December 1944

Cpl LeHenry L. Miller
- 19 December 1944

Cpl. James M. Self
- 19 December 1944

Cpl. Philip W. Skufoa
- 19 December 1944

Tech 5 Joseph S. Bowen
- 19 December 1944

Tech 5 Joseph J. Cipov
- 19 December 1944

Tech 5 Robert H. Dawson
- 19 December 1944

Tech 5 Horace Gibson
- 19 December 1944

Tech 5 Forrest A. Malone
- 19 December 1944

Tech 5 Kenneth L. Nelson
- 19 December 1944

Tech 5 Chester A. Roelofs
- 19 December 1944

Tech 5 Gustav W. Stenberg
- 19 December 1944

Tech 5 August C. Smith
- 19 December 1944

Pfc. William W. Farmer
- 19 December 1944

Pfc. Pedro Gutierrez, Jr.
- 19 December 1944

Pfc. Perry N. Harrison, Jr.
- 19 December 1944

Pfc. Alvin B. Huffman
- 19 December 1944

Pfc. James T. Johnson
- 19 December 1944

Pfc. James C. Keefe
- 19 December 1944

Pfc. Earl P. Mattison
- 19 December 1944

Pfc. David Segal
- 19 December 1944

Pfc. James E. Turner
- 19 December 1944

Pvt. Emil J. Bradac
- 19 December 1944

Pvt. Orville A. Brenden
- 19 December 1944

Pvt. Daily W. Colvin
- 19 December 1944

Pvt. William B. Dailey
- 19 December 1944

Pvt. Robert W. Hightower
- 19 December 1944

Pvt. Walter E. Lathe
- 19 December 1944

Pvt. Joseph W. Powell, Jr.
- 19 December 1944

Pvt. Calvin W. Smith
- 19 December 1944

Pvt. Robert E. Smith
- 19 December 1944

Pvt. Carl Wycoff
- 19 December 1944

Tech 5 Sylvester P. Taylor
– 20 December 1944

Pfc. Herbert L. Brasher
– 20 December 1944

Pfc. Lacy W. Brewer
– December 20 1944

Pvt. Sandy Mendieta
– 21 December 1944

1st Lt. Michael L. Runey, Jr.
– 22 December 1944

Sgt. William J. Mueller
– 22 December 1944

Tech 4 John B. Hartlage
– 22 December 1944

Cpl Gerald S. Ballard
– 22 December 1944

Cpl Frank Spayd, Jr.
– 22 December 1944

Tech 5 Olin L. Arnold
– 22 December 1944

Tech 5 – Eustace P. Dixon
– 22 December 1944

Pfc. Leonard H. Johnson
– 22 December 1944

Pvt. Lester L. Reed
– 22 December 1944

Pvt. Floyd K. Roberts
– 22 December 1944

Pvt. Lovic C. Cauthen
– 23 December 1944

Awards Presented December 1944

Date of Action

1st Lt. Carl M. Johnstone, Jr. – Oak Leaf Cluster to Silver Star
– 6 and 7 June 1944

Service

Major Starling W. Wood, Jr. – Bronze Star
– 2 Nov. 1942 – 5 June 1944

By the middle of January 1945, though not fully equipped, the Battalion began to look more like a fighting artillery unit. There was a relieved sigh all over the 58th, and people quit worrying about things like platoons, bayonet charges, and walking through Germany. On the morning of 18 January, we rode off to the wars once more, leaving much sorrow in Fontenoille.

The first position reached was at Foy, Belgium. The Battalion was now attached to the 11th Armored Division and the 174th Field Artillery Group, tactically reinforcing the fires of Division Artillery. This position was maintained through 27 January, with very little firing done after the first few days.

On 21 January a half-track was lost to an enemy mine, and the driver wounded and evacuated. The weather was bitter cold through this period, and it was rough going on the boys and the guns. On 25 January the Battalion was relieved of attachment to the 11th Armored Division and 174th Field Artillery Group, and on the 27th was attached to the 4th Infantry Division, to reinforce the fires of the 42nd Field Artillery Battalion in direct support of the 12th Infantry. On 28 January the Battalion moved to a position near Durler-Hof, Belgium, traveling 27 miles. The Battalion remained in position, firing on 29 January. March order was received again on 30 January, and the Battalion moved to Brecht, Belgium, a short move this time.

On 31 January the 58th sent three observers to join the 12th Infantry Regiment. The Battalion fired several T.O.T.s throughout the day. We received a few rounds of enemy artillery fire near one of our battery positions late in the afternoon.

Losses in action January 1945

Pfc. Harold J. Lockwood
– 26 January 1945

Awards presented January 1945

Date of Act

1st Lt. Edward F. Kasold – Bronze Star
– 17-19 November 1944

Captain Kenneth E. Beitler – Distinguished
Service Cross
– 6 June 1944

1st Lt. Vincent E. Baker – Bronze Star
– 11 October 1944

2nd Lt. John R. Jackson, Jr. – Bronze Star
– 13 September 1944

S/Sgt Mayo S. Herrington – Bronze Star
– 18 November 1944

S/Sgt Robert H. Wise – Bronze Star
– 17-18 November 1944

Cpl. Delbert W. Green – Bronze Star
– 17-18 November 1944

Cpl. Rodman E. Miller – Bronze Star
– 7 November 1944

Pvt. William B. Dailey – Bronze Star
– 7 November 1944

S/Sgt. Melvin W. Padgett – Soldier's Medal
– 17 November 1944

Sgt. Evan W. Holyfield – Soldier's Medal
– 17 November 1944

Pfc. Steve R. Hanson – Soldier's Medal
– 17 November 1944

2nd Lt. Johnny J. Horvath – Silver Star
– 20-28 December 1944

Sgt. James L. Caddell – Silver Star
– 20 December 1944

Sgt. Evan B. Lum – Silver Star
– 19-20 December 1944

Sgt. Walter J. Abramski – Silver Star
– 19 December 1944

The Battalion remained in Brecht until 2 February. The cold, snow and the swollen rivers were slowing the infantry advance considerably. Our three observers continued operating with the 2nd Battalion, 12th Infantry Division as, plagued by visions of punching away making different holes in the Siegfried Line for the next six months, we once more made the entry into Germany.

GERMANY

The first position occupied on this second pilgrimage to the home of the Harrenvolk was at Orb, Germany. One gun displaced forward at 0900 and reached the new position at 1100. Air OP registered this gun on the Base Point. The rest of the Battalion was in position by 1700, and we fired several harassing missions throughout the night.

On 4 February the Battalion moved to new positions south of Schonberg. Germany looked even worse the second time the Battalion entered it than it had the first time. The snows were just departing, the towns all looked dirty, and everywhere there was mud. Of course, there had been a certain amount of shells and bombs tossed around. And it doesn't do any town a bit of good to have innumerable tanks milling around in the front yards.

The Battalion was now in direct support of the 4th Reconnaissance Troop, whose mission was to protect the left flank of the 4th Infantry Division and to maintain contact with the 87th Infantry Division on the left (north) flank. The Mission of the 4th Infantry Division was to clear the enemy

111

in zone and to capture the town of Brand-schied. One of the 58th's FOs was sent with the leading elements to adjust fire.

At 1600 hours on 4 February the Battalion received verbal orders that we were no longer in direct support of the 4th Reconnaissance Troop, but were now in general support of the 4th Infantry Division, reinforcing the fires of the 29th Field Artillery Battalion, who were in direct support of the 8th Infantry Regiment. The mission remained the same.

The Battalion moved again on 6 February to the vicinity of Hatenfield. In the afternoon a few rounds of mortar came bucketing into the forward area, and a concentration of 75mm fire, but no damage or casualties resulted. The Battalion remained in this position, waiting for the high ground to the east to be cleared. On the 9th, the 58th moved to a position on the fringe of a woods near Waschied. The new area had not been cleared of mines, and one 1/4-ton vehicle tripped over one. Both occupants of the vehicle were injured, and the driver was evacuated. The sector through here was very heavily mined, and precautions had to be taken in every area to prevent similar accidents.

The Battalion remained in the woods through the remainder of the month of February, living in pyramidal tents. On 11 February General Blakely, Commanding General, 4th Infantry Division, visited the Battalion and presented individual awards to several of the men and officers. The 58th received verbal orders at 1800 attaching us to the 402nd Field Artillery Group.

On 12 February we received written orders that we were not attached to the 402nd, and our mission remained the same. A small enemy counter-attack developed northeast of Hermespand at 0545 hours. This attack, which consisted of two enemy tanks, one SP gun, and approximately two companies of infantry, was successfully broken up by artillery fire, causing the tanks to withdraw and inflicting casualties among the enemy infantry.

The mission of the Division at this point was to clear all ground west of the Prum River and to capture the town of Prum. On 15 February the enemy dumped a large concentration of rocks on the east edge of the Battalion position, but no casualties or damages were sustained. Again on the 17th a large concentration of light artillery fell just short of the battery positions, and 10 rounds landed in C Battery's area.

Two preparations were fired on the morning of 18 February to assist the 90th Division in their attack to the south. Their mission was to drive south, clearing the enemy west of the Prum River to Corps boundary. On the 20th the Battalion fired several observed missions against enemy infantry, and reports from captured prisoners indicated the results were excellent. One of the prisoners stated that 30 men out of the company's 70 were killed or wounded by our artillery fire. One of the Battalion's

113

FOs, 1st Lt. Collins, was mortally wounded on 23 February while adjusting fire.

On 27 February the Battalion fired preparations to support the 87th Infantry Division attack. The attack proceeded slowly but without much opposition. We received plans for an attack by the 4th Infantry Division, which was scheduled for 28 February. Their mission was to capture Budesheim and to seize the high ground west of the Kyll and Oos Rivers in zone. Our observers were sent to the 70th Tank Battalion to support them.

The 2nd and 3rd Battalions of the 8th Infantry Regiment attacked across the Prum River at H-Hour, 0515, and met heavy resistance from dug-in positions on the east side. Also many A.T. and A.P. mines were encountered. Troops took their objectives but had not completely secured them. Late in the afternoon Air OP fired on two enemy batteries and two rocket positions. Good effect was observed on all targets.

Losses in action February 1945

T/5 Woodrow P. Cato
– 9 February 1945

S/Sgt Robert H. Wise
– 11 February 1945

1st Lt. Clifford V. Collins
– 23 February 1945

Awards presented February 1945

Date of action

Sgt. Charlie S. Harrelson – Silver Star
– 19 December 1944

T/4 Theron C. Derrick – Bronze Star
– 22 December 1944

Pfc. Herbert L. Brasher – Silver Star
– 19 December 1944

Capt. Jack P. Partlow – Bronze Star
– 20 December 1944

The advance of the infantry continued on 1-2 March, and the 70th Tank Battalion, accompanied by our observer, moved into Willerath on the 2nd. The Battalion fired missions against enemy rockets, with adjustments of fire being made by Air OP. The tanks met more opposition on the 3rd from antitank guns on the flanks. The Battalion fired smoke missions, but high wind deprived these of effectiveness.

The Battalion was detached from the 402nd Field Artillery Group and reattached to the 333rd Group, in general support of the 11th Armored Division, and ordered to displace in the vicinity of Willerath. One battery was displaced forward at once, and the remaining two supported the attack of the 4th Infantry Division. The complete

Battalion closed in the new area near Willerath at noon on 3 March.

On the morning of 4 March the Battalion Commander, Lt. Colonel Paton, while reconnoitering for a new position, was killed when a mine blew up the "Weasel" in which he was riding. He was the second Battalion Commander of the 58th killed in action. The reconnaissance offer, 1st Lt. William M. Donnelly, and the driver were seriously injured, and both were evacuated. Major Starling W. Wood, Jr. assumed command of the Battalion.

The 11th Armored Division had jumped off in the morning, and the Battalion supported the attack with several T.O.T.s on heavily defended enemy positions. On the morning of 5 March the Battalion moved to a position in the vicinity of Budesheim. The road was shelled quite heavily just as the end of the Battalion column was moving into Budesheim, but no damage or casualties resulted. On 6 March, the Battalion remained in position, supporting the crossing of the Kyll River with preparation and smoke missions.

The Battalion was relieved from attachment to the 333rd Group on the 7th, and attached to CCA of the 11th Armored Division. We moved to Hinterhausen to support another crossing of the Kyll River at Lessingen. The Battalion joined the column of CCA and was off on an old time armored drive.

The 11th Armored objective was Kelberg-Mayen, and the capture of Andernach on the Rhine. It was definitely a "rat-race", until just short of Boxberg an anti-tank gun let part of the column through and then proceeded to knock out five vehicles in the middle of the column. East of Dreis Bonn, the Battalion went into position and fired harassing fires throughout the night. The anti-tank gun was cleaned out by infantry.

On the morning of 8 March the column started rolling again, moving through Boxberg, Kelberg, Boos, Mayen, Hausen, and Ochtendung. The Battalion went into position southwest of Plaidt at 0200 hours the morning of 9 March. During the march, many groups of prisoners of war were observed going to the rear without guards. Everyone was just too occupied to bother with going through the formalities of taking them prisoner. It must be rather embarrassing to try to give yourself up, and find no takers.

Our leading elements jumped off at 1200 hours on 9 March and were in Andernach by mid-afternoon. There was some fire from bazookas and small arms from across the river. One of our FOs, Lt. Johnny J. Horvath, fired on the enemy anti-tank guns that were holding up our tanks and successfully neutralized them.

The Battalion remained in position in the vicinity of Plaidt until 15 March. A few missions were fired against enemy infantry and tanks across the Rhine River. The weather had cleared considerably now, and our planes were busier than beavers all day long.

We heard a rumor that we were no longer attached to the 11th Armored Division, and were going with the 87th Infantry Division. Major Wood visited Division Artillery and was given a goose egg to occupy, which was in an exposed position. The 11th Armored Division said we would move on their order.

On 15 March the Battalion moved to another position at Mulheim. The 58th fired two observed missions plus several H & I in support of the 87th Division's crossing of the Moselle River. The Battalion was now firing in reinforcement of the 334th Field Artillery Battalion, in direct support of the 335th Infantry Regiment. The 58th fired two more observed missions and 4 T.O.T.s on 16 March. We received verbal orders that we're now attached to the 11th Armored Division, who were detached to 12 Corps, and then reattached to the 33rd Field Armored Brigade.

On 17 March the Battalion left its position at 0010 and arrived at a new position near Schmidt at 0730. We remained near Schmidt until 1600, and then traveled through Driesch, Lutzerath, Kenngus, Bad Bertrich and Alf, and crossed the Moselle at Bullay around 1900. One enlisted man and one half-track were lost here, the half-track temporarily. Rifle fire through the windshield killed the driver.

On 18 March, FO Lt. Vincent E. Baker adjusted his battery to silence an enemy anti-tank gun and drive it back into the town of Kirchberg. He then directed the 58th's fire upon the town and removed a serious threat to the column and to the advancing infantry. The Battalion continued moving, past Eirchberg and into a position southeast of Dickensheid., where we remained throughout the night.

In the morning the march was resumed, and a position occupied near Hainzenberg. This position was in an extremely peaceful-looking valley. The sun was shining, and the sky was as blue as any spring sky was ever blue. Everything was quiet, and the men were standing around, more or less "batting the breeze."

Then, with no warning, rockets began falling in the area. Approximately 30 rockets fell in a space of a few seconds. The majority of them landed on the hill slope at the edge of B Battery, but several landed directly in the battery area.

As near as anyone could estimate there were about 30 of them, though practically everyone lost count after the first round. Several of B Battery's men had been caught out in the open; 3 enlisted men were killed, 3 B Battery officers wounded and evacuated, and 8 enlisted men wounded and evacuated. At no time in its history was the 58th hit so hard in such a short space of time. No more rockets came in after that one salvo.

Shortly before midnight, the Battalion moved again, traveling all night. On the other side of Heinzenberg a small tank battle appeared to be going on to the right of the column, with a few tracers looping up toward the town. The Battalion crossed the Nahe River near Simmern on 20 March, and went into position at 0515 hours west of Meddersheim. At 0930 hours the Battalion was on the move again, moving south through Meisensheim, Callbach, Gangloff, Ransweiler, Schonborn, and Katzenbach. We moved into position at 1330 and remained there for several hours without firing. Then we started moving once again, through Marienthal, Dannegeis and Dreisen, closing into position at 2100 at Albisheim. A Battery, which was leading today, fired one direct fire mission against enemy entrenched in and around a building, with excellent effect.

The weather was clear and warm now, and roads were becoming dusty again. In armored drives, the roads were either swamps or dustbowls. There seemed to be no in-between.

The town of Albisheim had just recently been worked over pretty thoroughly by artillery and planes. Buildings smoked and burned as the column rumbled through the debris-laden streets, and civilians still stood with that stunned look on their faces. German ack-ack guns poked their noses out of the wreckage. Apparently they had run them right into the back yards of houses and set them up.

The afternoon of 21 March the Battalion moved again, attached to the 183rd Group. The next position was near Hahnheim. Through here, there was increasing evidence of activity on the part of the German Chamber of Commerce. Buildings were plastered with signs, such as "See Germany and Die!" and "Onward, Slaves of Moscow."

The following morning the 58th rolled on to Winterheim, where it was given the mission of supporting the 42nd Cavalry Squadron, holding defensive positions on the Rhine. A few observed missions were fired the morning of the 23rd, and a preparation was fired to support the crossing of the river by the 5th Infantry Division. In the afternoon of the 23rd the Battalion moved to an assembly area with orders to move from this point and cross the Rhine in the vicinity of Oppenheim.

The Battalion moved out of the assembly area shortly before midnight, moving slowly toward Oppenheim. The Luftwaffe was out and was constantly strafing the bridge and its approaches. As the Battalion crawled through Oppenheim, the Germans opened up with interdicting fire with 105s from across the river. The slow pace, the whistle of shells, and the threat of strafing formed a combination a bit hard on the nerves. The lead vehicles pussy-footed up to the pontoon bridge about 0300 hours and, with one man walking in front of each vehicle as a guide, crossed the historic Rhine. The map doesn't show it, but at this point, the Rhine was slightly wider than the English Channel—or it was that night, anyway. There was no strafing while the

Battalion crossed, but the 105s kept it from being too dull.

On the east side of the Rhine the Battalion went into position southeast of Geinsheim at 0630 hours. Enemy planes were overhead early in the morning, but kept pretty well up. The Battalion fired several missions for the 183rd Group and the 90th Infantry Division. Another move was made in the afternoon to Dornheim, where the Battalion remained for the night. The Luftwaffe was overhead again during the night, bombing and strafing, but no damage was sustained by the Battalion.

The morning of the 25th a move was made to Buttleborn. Here the mission of the 58th was changed to direct support of the 6th Armored Division while they crossed the Rhine that day. Three moves were made that afternoon before the Battalion joined with the 86th Reconnaissance Battalion. Then a move was made to a fourth assembly area. Two of the Battalion FOs were sent out with each of the 86th's tank forces. Our mission was direct support of the 86th, whose mission was to clean out in between CCA and CCB.

On the night of 25 March, shortly after dusk, enemy planes strung out an exceptionally heavy concentration of flares over the bridges near Oppenheim. There were at least 100 flares in the concentration, and the surrounding area was lit up for miles. Service Battery was approaching the bridge to cross at this time. A heavy bombing attack was anticipated on the bridges, but never materialized, and it was later learned that our night-fighters had intercepted the enemy bombers short of the target and forced them to jettison their bomb loads. Smoke generators had been placed in operation now, and a heavy artificial fog covered the bridges and approaches day and night.

On 26 March the column moved out in the morning. Our mission was to protect the Division's right flank and to make contact with the 4th Armored Division on our right. Also to drive into the town of Offenbach and seize any bridges across the Main River. We drove through Morfelden, Langen, Dudenhofen, and Jugenheim. Just short of Heussenstamm, the head of the column ran into fire from an 88mm AA gun and a little 20mm fire. Our Air OP adjusted fire on this target and neutralized the guns, enabling the column to advance. The Battalion closed into an assembly area north of Neu Isenberg to await completion of a bridge across the Main River at Frankfurt.

The Battalion remained in position until the afternoon of 27 March, when a new position was taken, where the Battalion reinforced the fires of the 11th Armored Division Artillery and fired T.O.T.s throughout the night. This firing continued on the morning of 28 March, when the Battalion was informed it was now reinforcing the fires of CCA in direct support of Task Force Davall (confusing, isn't it?). In column with CCA the Battalion

123

was pulled off the road and coiled around the bridge site on the Main River. This was a particularly hot spot to coil, inasmuch as enemy planes were attacking the bridge.

Early in the morning of the 29th the Battalion moved to join the 11th Armored Division. We crossed the Main River at 0910 hours and then moved through Hanau. One battery moved with the advance elements of the column, and there was always one battery in position during the march.

The Battalion was now in direct support of Task Force Brady, in general support of CCA, and attached to the 183rd Group. (Are there any questions?) The Battalion occupied positions near Langenselbold, as the task force had run into heavy opposition just short of Rothensbergen. The 58th lost 1 half-track, and one enlisted man from one of the FO parties was wounded. There was also slight shelling in the Battalion area throughout the night.

The morning of 30 March the Battalion fired preparation for advance of the task force just before the jump-off. Two towns were taken, but the advance was held up short of Glenhausen. The Battalion moved into position to fire all day at observed targets. Good results were obtained. Three AT guns, and two personnel carriers were knocked out, and heavy personnel casualties inflicted on the enemy. The Battalion also fired a smoke preparation to enable our infantry, who were pinned down by A/T fire and small arms fire, to withdraw from their exposed position. Major General Maurice Rose, Commanding Officer of the 3rd Armored Division, for whom the 58th shot artillery, was killed today near Paderborn.

The morning of 31 March the column was moving again, bypassing the resistance at Glenhausen, which would be taken care of by the 26th Infantry Division, and finally going into position in the evening at Kressenbach. The speed with which the columns were moving now left pockets of enemy resistance to be cleaned up by the infantry. Occasionally this by-passed resistance made it rough on the supply column coming up behind, and the roads down which the spearheads advanced were frequently closed to supply trains by enemy action.

Losses in action March 1945

Pfc. Harry R. McSparrin
– 4 March 1945

Lt. Colonel Walter J. Paton
– 4 March 1945

1st Lt. William M. Donnelly
– 4 March 1945

Pfc. John J. Quinn
– 4 March 1945

T/5 Paul B. English
– 17 March 1945

1st. Lt. Edward J. Kasold
– 19 March 1945

1st. Lt. Richard J. Duggan
– 19 March 1945

2nd Lt. Stanley C. Dukiet
– 19 March 1945

Sgt. Albert A. Kneer
– 19 March 1945

Cpl. Charles W. Whitcomb
– 19 March 1945

Pfc. Gonsolio Vincent
– 19 March 1945

T/5 Irvin I. Cammer
– 19 March 1945

Cpl. Steve Brier
– 19 March 1945

Cpl. John E. Clinard
– 19 March 1945

Pvt. Edward A. Diamos
– 19 March 1945

T/5 Clyde C. Freeman
– 19 March 1945

T/5 Vester A. Logan
– 19 March 1945

Pfc. Gerald W. Maschmann
– 19 March 1945

T/4 Key W. Moore
– 19 March 1945

Cpl. James E. Stephens
– 29 March 1945

Awards presented March 1945

Date of action

1st. Lt. Edward J. Kasold – Bronze Star
– 19-22 December 1944

T/4 Rocco Paulercio – Bronze Star
– 19 December 1944

Cpl. Frank Spayd, Jr. – Bronze Star
– 19 December 1944

Sgt. Charles P. Hunt – Bronze Star
– 19 December 1944

Cpl. Severt Gilderhus – Bronze Star
– 19 December 1944

Major Kenneth E. Beitler – Croix de Guerre
– 6 June-1 September 1944

2nd Lt. James C. Hipp – Croix de Guerre – 6 June-1 September 1944

The Battalion moved again on 1 April, passing through Waliroth, Weideau, Hauswurz, and through the woods to Grossenlider to Shlitz. The Division attacked north, just east of Fulda, with the objective of Kranichfeld. CCB led the attack followed by CCA passing through Michelsrombach, Nust and Morles. Just short of Obernust, CCB took the north route and CCA the southern route. CCA passed through Hilders, and the Battalion took a position shortly after midnight northeast of Frenkenheim.

In the morning the march continued, passing through Honneberg. A defensive position was taken just short of the Werra River. Resistance was light, and very little artillery fire was called for.

The morning of 3 April the Battalion crossed the Werra River and moved up towards Suhl, where small-arms and anti-tank fire was encountered, coming from the vicinity of the city. (Suhl had been quite a small arms manufacturing center; several pistol, rifle and shot-gun factories were located there.) In the vicinity of Marisfeld, artillery and rocket fire fell on the Battalion, but no damage or casualties were sustained. Positions were occupied at Bischofrod, with one battery forward to cover the northern approach to Suhl.

The Battalion remained in position on

the 4th, firing against enemy infantry. Suhl had been cleared, but well dug-in enemy positions on the outer fringes of the town were offering resistance. The Battalion's fire inflicted severe casualties on the enemy infantry, and the high ground east of the town was cleared. The Battalion moved to a new position in the evening.

The morning of 5 April the Battalion moved again to Goldlauter, and fired close support missions for reconnaissance troops and infantry. By nightfall everything had pulled back but one platoon of infantry and the Battalion. The next day the Battalion was ordered to return to its former positions at Suhl, and was relieved from 183rd Group and XII Corps and attached to XX Corps and reattached to the 416th Field Artillery Group.

The morning of 7 April the complete Battalion departed from Suhl, moving to an assembly area in the vicinity of Unter-suhl. We traveled all day, part of the time on the Super-Highway (Autobahn). These highways were constructed to expedite military movement within the Reich. They were doing just that now, but not exactly as planned in the New Order.

The following day, 8 April, the Battalion made a short move to Markershausen, where maintenance was begun on vehicles. On 9 April the Battalion was relieved of attachment to the 416th Field Artillery Group, and of all things, attached to the 5th Artillery Group, the Battalion's old group from Africa and Sicily. Many old acquaintances were renewed. The 62nd and 65th Armored Field Artillery Battalions were no longer in the group, but the Group Headquarters itself had many of the old personnel.

On 10 April the Battalion departed again, to Markershausen and Boilstadt, where an assembly area was reached at noon. The Battalion was given the mission of reinforcing the fires of the 94th Field Artillery Battalion, in direct support of CCR, 4th Armored Division, whose objective was Dresden. The 58th moved out on 11 April and, traveling over dusty roads, went into firing position late at night south of Erfurt, firing one harassing mission on Weimar.

On the 12th the Battalion moved up with the column and occupied positions east of Weimar at Taubach. The following day they crossed the Saale River, occupied positions at Nennewitz, and moved on again to Rudersdorf in the evening. Moving out on the Autobahn again the next morning, positions were occupied at or near Dennherritz. Some firing was done on the town, and enemy personnel cleared out.

The next morning the Battalion departed to Remse. At Remse the Battalion worked on its broken-down vehicles in an effort to ready them for combat. Since the Battalion had been re-equipped in Fontenoille the first part of January, there had been no opportunity to perform more than the most perfunctory maintenance on the vehicles, and things were beginning to fall apart. The Battalion had reached a point where it was just about unable to fight, unless the gun crews carried the M-7s on their backs.

The Battalion remained at Remse until the morning of 19 April. Word was received that XX Corps was moving to the southern flank of the Third Army front, and the 58th, no longer attached to 4th Armored Division, took off for an area north of Bamberg.

The first half of the move was back down the Autobahn. Apparently the move coincided with that of a couple more artillery battalions and an engineer outfit. M-7s, Long-Toms, 6x6s and tanks went milling down the broad highway four abreast. It was quite a show—like a mechanized adaptation of the opening of the Cherokee Strip. A column of German prisoners in 6x6s got strung out in the parade also, just to add to the general confusion. The Luftwaffe could really have raised hell on that road, but at that stage of the game the Luftwaffe wasn't raising much hell anywhere. The column closed into a position at Birkhach, after traveling 156 miles. Two tanks and several M-7s were left strung along the road.

On the afternoon of 21 April the Battalion was relieved of attachment to 5th Group. The mission of the Battalion was now to guard lines of communication. That portion of the Battalion which was still mobile was dispatched to the city of Bamberg to take up these duties.

On 24 April the Battalion received orders to relieve the 80th Infantry Division in the town of Nurnberg. Battalion Headquarters, Headquarters Battery and B Battery closed into Nurnberg at 1930 hours. A and C Batteries proceeded to Nurnberg, closing into position at 2000 hours.

Our mission was the security of Nurnberg. We were unable to place guards on the number of places or have as many patrols as the 80th Infantry Division had, but by putting out skeleton guard we were able to cover the town fairly effectively. Our main interest was displaced persons camps, food warehouses, clothing warehouses, and firearms. Our main trouble came from displaced persons. It was a job to keep the looting under control.

Losses in action April 1945

None

Awards presented April 1945

Date of act

1st Lt. Edmund M. Hamlin – Silver Star – 19 December 1944

Major Starling W. Wood, Jr. – Oak Leaf Cluster to Bronze Star – 16-23 March 1945

The Battalion received vocal orders on 1 May that B Battery was to move to guard a gas dump at Leierndorf, and A Battery and the Battalion CP were to move to Obtraubling. Battery C would remain in Nurnberg to guard various installations.

AUSTRIA

On 5 May the Battalion took over security guard duties at Simbach on Inn, Austria, just across the river from Braunau—the birthplace of Adolph Hitler. The 58th had finally run him down. The Battalion took over the Military Government in Simbach on Inn and Braunau. When C Battery arrived from Nurnberg they took over security guard of the cities.

On 7 May C Battery departed for Wels, Austria, to take over various installations when the 71st Infantry Division left. C Battery patrolled from Reid to Lambach to Wels. Half of B Battery continued to guard the gas dump at Leierndorf while the other half moved to Lambach and patrolled from Strasswalchen to Vocklabruck back to Lambach. Headquarters Battery patrolled from Braunau to Reid, and also from Braunau to Strasswalchen. Major Starling W. Wood, Battalion Commander, appointed 1st Lt. Vincent E. Baker to serve as mayor of Reid, Austria, under the auspices of the American Military Government.

Losses in action May 1945

None

Awards presented May 1945

Date of action

Cpl Donal E. Fee – Bronze Star
– 19 December 1944

T/5 Charlton A. Harp – Bronze Star
– 19 December 1944

1st Lt.Vincent E. Baker – Oak Leaf Cluster
to Bronze Star
–18 March 1945

VE-DAY - 8 MAY, 1945

On 8 May the Germans surrendered unconditionally. The Battalion's position and mission remained the same until 25 July, when the following officers of the 58th Armored Field Artillery Battalion were transferred to the 10th Armored Division to begin the long-awaited trip home to the United States of America:

Capt. Lewis B. Fisher
Capt. Kenneth M. Frawley
Capt. Nathaniel B. Smith
Capt. Donald D. Moore
1st Lt. Vincent E. Baker
1st Lt. Robert A. Davis
1st Lt. Floyd E. Dickerson
1st Lt. Richard J. Duggan
1st Lt. Wayne J. Fye
1st Lt. Eugene Glowa
1st Lt. Johnny J. Horvath
1st Lt. Benjamin J. Noyes
1st Lt. Leroy C. Stevens, Jr.
1st Lt. Tyssul G. Thomas
2nd Lt. James C. Hipp
2nd Lt. Jack James
CWO John T. Eastburn

FINIS

This more or less brings the combat history of the 58th Armored Field Artillery Battalion up to date. Looking back, this history seems chiefly a record of travel, attachments and assignments—some 15,000 miles of sea and land travel; 31 months overseas; 12 foreign countries; 417 days of combat; attachment or assignment to 4 Armies, 7 Corps, and 19 Divisions. In fact, the Battalion has been attached and detached more often than Gypsy Rose Lee's garter belt.

It doesn't tell very well about the endless grinding through swirling dust; the everlasting cry of "March Order" in the middle of the night; the mixed feeling of futility and fear huddled in a slit trench during counter-battery; the sordid monotony of death by the roadside, of blasted homes, and burned and twisted metal. And it doesn't tell of the close, warm relationship of men who have for two-and-one-half years lived, sweated, laughed, feared, and played together in a strange land, and in the uncertain shadow of pain, destruction, and death.

Thus we close, at least temporarily, the story of the Five-Eight and its officers and men—a pretty nice bunch of guys. Some of them drank a bit too much. And when the chips were down, some of them died in the lemon groves of Sicily, on the beach at Normandy, and in the frozen fields around Bastogne.

General Eisenhower said a few days ago that troops who fought in Africa and the ETO will not fight in the Pacific. This is accepted in the Battalion with polite enthusiasm. Next week the 58th will probably be in direct support of the Philippine Scouts!

END NOTE

This combat history of the 58th Armored Field Artillery Battalion was compiled from after action reports, transit notebooks, written histories and oral interviews with both officers and enlisted men, and contemporaneous records of rounds fired from 6 June 1944 through 31 October 1944.

An attractive bronze plaque with the design of the 58th's shoulder patch and the statement, "The 58th Armored Field Artillery Battalion landed here 6 June 1944" was made in the United States and shipped to France. Mounted on a stone base made for it, it is in place at the Vierville draw at the west end of Omaha Beach where the 58th landed. The French mayor of Vierville arranged for the marker to be dedicated on 6 June, 1984, the 40th anniversary of the D-Day landings. The ceremony was followed by a vin d'honeur for the little unit which never numbered more than 800 and was commanded by a lieutenant colonel. Eight generals, the 1st Division color guard, a 180-piece band, and a large throng of American and French people heard the address of William M. Donnelly, CPT, Retired, as he dedicated the marker to Lt. Colonel Bernard William McQuade. Following is a quotation from that speech:

139

"Our beloved commanding officer had led us this far; he got no further, dying in shallow water by machine gun fire while attempting to drag his wounded sergeant to safety. Lieutenant Colonel Bernard W. McQuade was his name. Tough, wiry, acid-tongued (he could really take it off in painful strips), brilliant, brave, impatient, tireless, a perfectionist, and fair. The officers knew it; the men knew it. We all adored him."

The D-Day experiences of then-Lt. Vincent E. Baker as well as other soldiers who landed on the beach can be read at the Omaha Beach Memorial:

http://www.omaha-beach-memorial. org - Click on Eyewitnesses

ABOUT THE EDITOR

Fran Baker is the author of 18 bestselling novels as well as numerous articles, book reviews, and opinion pieces.

Capt. (Ret.) Vincent E. Baker, her late husband, graduated from Officer Candidate School at Fort Sill, Oklahoma on 18 March 1943, before joining the 58th Armored Field Artillery Battalion in England on 3 January 1944. He landed on Omaha Beach on 6 June 1944 with the 58th AFABn and fought as a forward artillery observer from D-Day through VE-Day, when he was appointed military mayor of Reid, Austria. Vincent was wounded in combat five times, was awarded two Purple Hearts, two Bronze Stars with the V device for Valor, the Bronze Initial Landing Arrowhead and five campaign stars. He was field-promoted from 2nd Lieutenant to 1st. In 2014 Vincent was appointed a Knight in the French Legion of Honor for his service in Normandy and Northern France. Fran's family story, *Once A Warrior*, is a fictional account of her husband's combat experiences with a little romance thrown into the mix. It's available worldwide in eBook format, and a limited number of print copies are available with free US shipping via her website.

Fran invites readers to visit her website at *www.FranBaker.com*.

www.ingramcontent.com/pod-product-compliance
Lightning Source LLC
Chambersburg PA
CBHW071500070426
42452CB00041B/1972